Lecture Notes in Civil Engineering

Volume 439

Lecture Notes in Civil Engineering (LNCE) publishes the latest developments in Civil Engineering—quickly, informally and in top quality. Though original research reported in proceedings and post-proceedings represents the core of LNCE, edited volumes of exceptionally high quality and interest may also be considered for publication. Volumes published in LNCE embrace all aspects and subfields of, as well as new challenges in, Civil Engineering. Topics in the series include:

- Construction and Structural Mechanics
- Building Materials
- Concrete, Steel and Timber Structures
- Geotechnical Engineering
- Earthquake Engineering
- Coastal Engineering
- Ocean and Offshore Engineering; Ships and Floating Structures
- Hydraulics, Hydrology and Water Resources Engineering
- Environmental Engineering and Sustainability
- Structural Health and Monitoring
- Surveying and Geographical Information Systems
- Indoor Environments
- Transportation and Traffic
- Risk Analysis
- Safety and Security

To submit a proposal or request further information, please contact the appropriate Springer Editor:

- Pierpaolo Riva at pierpaolo.riva@springer.com (Europe and Americas);
- Swati Meherishi at swati.meherishi@springer.com (Asia—except China, Australia, and New Zealand);
- Wayne Hu at wayne.hu@springer.com (China).

All books in the series now indexed by Scopus and EI Compendex database!

Sumit Saxena · Shobha Shukla ·
Prasanna Kumar S. Mural
Editors

Emerging Materials and Technologies in Water Remediation and Sensing

Proceedings of ICWT 2022

Springer

Editors
Sumit Saxena
Department of Metallurgical Engineering and Materials Science
Indian Institute of Technology Bombay
Mumbai, Maharashtra, India

Shobha Shukla
Department of Metallurgical Engineering and Materials Science
Indian Institute of Technology Bombay
Mumbai, Maharashtra, India

Prasanna Kumar S. Mural
Department of Metallurgical Engineering and Materials Science
Indian Institute of Technology Bombay
Mumbai, Maharashtra, India

ISSN 2366-2557 ISSN 2366-2565 (electronic)
Lecture Notes in Civil Engineering
ISBN 978-981-99-6761-2 ISBN 978-981-99-6762-9 (eBook)
https://doi.org/10.1007/978-981-99-6762-9

This Springer imprint is published by the registered company Springer Nature Singapore Pte Ltd.
The registered company address is: 152 Beach Road, #21-01/04 Gateway East, Singapore 189721, Singapore

Contents

A Comprehensive Overview of Vital Water Quality Parameters 1
Bishnu Kant Shukla, Mudasir Ahmad Teeli, Shubham Kumar Shukla, Rishabh Chandra, Nitish Bharti, and Utkarsh Singh

Surface Water Quality Modelling Using Water Quality Index (WQI) and Geographic Information System (GIS) on the Mahanadi Basin, Odisha 21
Abhijeet Das

Water for Food Drying: Geothermal Energy-Based Food Dryer 47
Namrata Bist, Roshni Kumari, Kriti Yadav, and Anirbid Sircar

Artificial Neural Network Modelling of Reverse Electrodialysis 59
Shruti Sen, Rudra Rath, Reddi Kamesh, and Anirban Roy

Development of Dairy Wastewater Treatment for Valuable Applications 73
Swati Sucharita Satpathy, Priti Chhanda Ojha, and Debabrata Pradhan

Application of Nanomaterials in Water Purification: A Thematic Review 97
Bishnu Kant Shukla, Rahul Dev Singh, Nitya Singh, Shivam Singh, Yash Kumar Burman, and Rahul Singh

Experimental Study on Green Synthesis of Copper Nanoparticles and Its Utilization for Degradation of Organic Dye 115
Sarvesh Datar, Shreyas Kadu, Pratiksha Khurpade, and Somnath Nandi

About the Editors

Dr. Sumit Saxena is Professor in the Department of Metallurgical Engineering and Materials Science at the Indian Institute of Technology Bombay, India. He obtained his doctoral degree at the New Jersey Institute of Technology, New York, USA. Subsequently, he worked as Postdoctoral Research Associate at the Department of Materials Science, University of Illinois—Urbana Champaign, Illinois, USA, and at the Department of Physics Harvard University, Cambridge, Massachusetts, USA. He is a fellow of the Royal Society of Chemistry, UK; Institute of Materials, Minerals and Mining, UK; and National Environmental Science Academy, India.

Dr. Shobha Shukla is Professor in the Department of Metallurgical Engineering and Materials Sciences at the Indian Institute of Technology, Bombay, India. She obtained her doctoral degree at the State University of New York/SUNY Buffalo, USA. Subsequently, she worked as Postdoctoral Fellow at the School of Engineering and Applied Sciences Harvard University, Cambridge, Massachusetts, USA. She is a senior member of the Optical Society of America and serves on the editorial board of Scientific Reports, Nanotechnology, and *Journal of Physics-Photonics.*

Dr. Prasanna Kumar S. Mural is Assistant Professor in the Department of Metallurgical Engineering and Materials Sciences at the Indian Institute of Technology, Bombay (IITB), India. He obtained his doctoral degree at the Indian Institute of Science, Bengaluru, India. He worked as Postdoctoral Fellow at the Council for Scientific and Industrial Research (CSIR) in Pretoria, South Africa. Subsequently, he joined the National Institute of Technology Calicut, India as Assistant Professor before joining IITB.

About the Editors

A Comprehensive Overview of Vital Water Quality Parameters

Bishnu Kant Shukla, Mudasir Ahmad Teeli, Shubham Kumar Shukla, Rishabh Chandra, Nitish Bharti, and Utkarsh Singh

Abstract Water is the furthermost significant component in the terrestrial foundation and regulating the climate. It is one of the chemicals that affects life the most significantly. Typically, the characteristics of water's physical, biological, chemical, and constitution are utilized to define it. Groundwater resources are being depleted and are becoming significantly and diversely contaminated as a result of rapid economic development and the unrestrained use of chemical pesticides and fertilizers in agriculture. In-place sewage systems, landfill leachate, wastewater treatment plant effluent, leaking sewers, hydraulic fracturing, gas stations, or excessive fertilizer use in agriculture are the usual causes of groundwater adulteration. Pollution can also be brought on by natural contaminants like fluoride and arsenic. Use of tainted groundwater exposes the entire public to illness or injury. Poor water quality can lead to a range of negative effects, including subpar goods, low earnings, and serious risks to human health. This document provides detailed information on underground water quality characteristics, including pH, temperature, biological oxygen demand, dissolved oxygen, alkalinity, chemical oxygen demand, nitrite, hardness, ammonia, nitrate, sulfate, and phosphate.

Keywords Water quality index (WQI) · Dissolved oxygen (D.O.) · Biochemical oxygen demand (BOD) · Alkalinity · Eutrophication · Heavy metal

1 Introduction

The lack of access to clean water is a global issue that affects all countries. The strong human desire to improve living standards has inevitably resulted in water pollution, with cumulative efforts and activities establishing as heavy industrial development and incessant urbanization, leading to progressive pollution of aquatic system [1–4].

B. K. Shukla (✉) · M. A. Teeli · S. K. Shukla · R. Chandra · N. Bharti · U. Singh
Department of Civil Engineering, JSS Academy of Technical Education, Noida 201301, India
e-mail: bishnukantshukla@gmail.com

S. Saxena et al. (eds.), *Emerging Materials and Technologies in Water Remediation and Sensing*, Lecture Notes in Civil Engineering 439,
https://doi.org/10.1007/978-981-99-6762-9_1

When it comes to drinking, irrigation of crops, waste disposal, transportation, recreation, irrigation of crops, and industrial production, humans are utterly dependent on renewable resources of freshwater. The quality and quantity of water that can be used to placate human necessities are already threatened in many parts of the biosphere due to rising water usage, growing population, climate change and a rising water usage, the gap between freshwater availability and demand will worsen over the next century. About 80% contamination of Indian water resources, according to estimates by WHO, is caused by inhabited refuse. Water system management issues could have a substantial impact on the supply of drinking water [5, 6]. Industrial waste is the main basis of water contamination. Wastewater from various industries is disposed of in the water without being properly treated [2]. Monitoring water quality is a vital component of environmental research and public health. It entails evaluating a range of physical, chemical, and biological factors to make sure that water is suitable for diverse uses, from supporting aquatic life to safe human consumption [7]. Temperature, turbidity, and color are some examples of the physical parameters. Aquatic creatures' biological activity and growth are directly influenced by temperature. Water clarity, or turbidity, can influence how aquatic plants synthesize their food and also reveal the presence of microbes. Minerals and organic debris that are dissolved in the water can be identified by looking at the color of the water. pH, conductivity, dissolved oxygen, hardness, and the presence of specific ions like nitrates, phosphates, and heavy metals are examples of chemical parameters. The solubility and toxicity of substances in the water can be influenced by the pH of the water, which is a measurement of its acidity or alkalinity [8]. Water's conductivity, a metric of how well it conducts electricity, is inversely proportional to its ion concentration. The life of aquatic species depends on the presence of dissolved oxygen. The usefulness of water for home and industrial usage can be influenced by its water hardness, which is determined by the amount of calcium and magnesium ions present. Heavy metals, phosphates, and nitrates can all be signs of pollution and provide health dangers [9].

A change in the underlying sediments' and aquifers' chemical makeup also affects the variation in water quality. Surface sources like canals, lakes, dams, and rivers provide around one-third of the world's drinking water needs. Water is used in some way or another by an organism's chemical and physiological processes. In numerous aspects of existence, water is crucial. In order to research numerous physicochemical water characteristics including B.O.D. (biochemical oxygen demand), WQI (water quality index), D.O. (dissolved oxygen), Eutrophication, heavy metal, alkalinity, etc., water samples from rivers and lakes provide an appropriate platform. Many scientists have occasionally studied the physicochemical properties of water bodies. The physicochemical features will also aid in the continuation of eco-biological impact research, the detection of pollution sources, and the beginning of remedial action procedures for polluted water bodies [10]. Numerous scholars in India have studied the physicochemical and biological properties of reservoirs and rivers. Since a variety of physical and chemical elements can combine to affect primary production levels, which in turn affect total biomass and trophy structure across the aquatic food web, water quality must be taken into account when evaluating wetland habitat.

Gaining a deeper understanding of these factors will help us manage our water resources more effectively, reduce risks, and guarantee the well-being and security of our communities and ecosystems. The function each of these parameters plays in determining the quality of the water, the variables that affect them, and the effects of their imbalance will all be covered in more detail in the sections that follow.

2 Analysis of Water Quality

Before using water for drinking, residential, industrial, or agricultural purposes, it must first be tested. Different biological, physical, and chemical factors must be used to test water. The only factors considered when selecting water testing parameters are the envisioned use of the water and the level to which its purity and quality are required [11–14]. Different kinds of microbiological, dissolving, suspended, floating, and bacteriological pollutants are present in water. Physical tests should be conducted to evaluate the substance's odor, color, temperature, pH, TDS, turbidity, and further physical characteristics. Chemical tests should be conducted to evaluate the substance's COD, BOD, alkalinity, hardness, DO, and other characteristics [15]. Water must be verified for heavy metal trace, organic, and metal contaminants like pesticide scum in order to obtain more and more pure, high-quality water. It goes without saying that potable water must satisfy all of the required parameters during tests and contain the necessary amount of minerals [16, 17]. Only in wealthy nations are all these requirements properly checked because there are very few organic and heavy metal pesticide contaminants in water, necessitating extremely analytical equipment and skilled labor. To monitor the water quality, a variety of physical–chemical parameters must be used.

2.1 Temperature Measurement

Using a digital thermometer, the temperature is determined by dipping the thermometer in the sample and recording the result [18]. Temperature is the degree of heat or cold in an aquatic organism's body, whether it is in water or on land. Despite having cold blood, fish experience body temperature changes in response to their environment. These changes have an impact on the fish's physiology and metabolism as well as its production. The microbiota's rate of metabolic activity as well as their rate of respiration both increase with temperature because of the rise in oxygen demand. It also results in increased ammonia levels in water and decreased oxygen solubility. However, with prolonged ice cover, chemicals such as carbon dioxide, methane, and hydrogen sulfide can build up to dangerous levels and affect the health of fish [19].

The accumulation of dangerous gases such carbon dioxide, methane, and hydrogen sulfide under varied lengths of ice cover is shown in Figs. 1 and 2. Figure 1

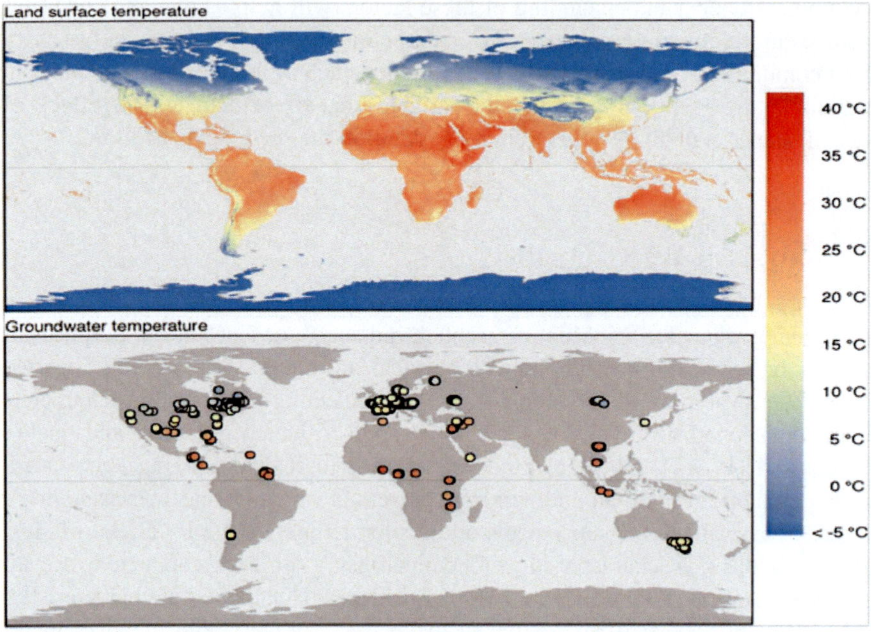

Fig. 1 Decadal average (2005–2014) of global shallow groundwater and land surface temperatures [19]

shows how chemical concentrations alter as ice cover duration lengthens. As seen in the image, there is a definite positive association between the quantity of these dangerous substances and the length of the ice cover. In particular, the data points point to a significant rise in chemical concentration following a specific critical period of ice cover. This suggests that a protracted ice cover may eventually cause a buildup of these compounds to levels that are harmful to aquatic life. Figure 2, on the other hand, depicts the condition of fish populations under various chemical concentrations brought on by various ice cover durations. The graph clearly shows that fish health drastically declines when chemical concentrations increase, a scenario brought on by prolonged ice cover. The need of preventing protracted ice cover scenarios is highlighted by this research in order to protect aquatic biodiversity and fish health.

2.2 pH Measurement

The pH scale, which determines how acidic or alkaline a solution is, fluctuates depending on the geology, contaminants, and climate of the area. For instance, locations with granite bedrock may have more acidic water than those with carbonate bedrock. Additionally, pollutants and dissolved gases can affect pH. Although drinking water is frequently supplemented with minerals and has a pH of 7, pure water

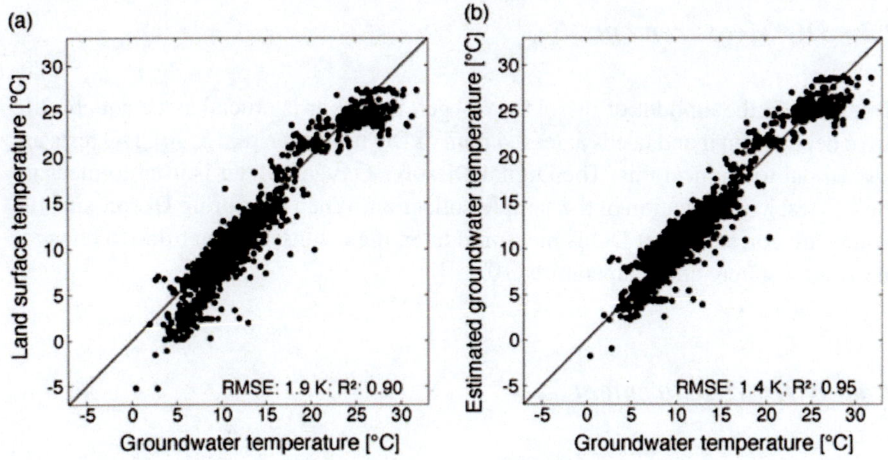

Fig. 2 **a** Correlation between land surface and groundwater temperatures; **b** Comparison of estimated and measured groundwater temperatures [19]

is inherently more alkaline. The World Health Organization recommends drinking water with a pH of 6.5–8.5. Beyond this range, water may provide health problems because extremely low pH can lead to corrosion and the release of metal ions from pipes, while extremely high pH can occasionally result in skin and eye irritation [20]. Due to the inclusion of minerals, bottled mineral water like Bisleri may have a pH slightly higher than 7. It's generally accepted that keeping a balanced pH in drinking water is essential for health, despite occasional claims to the contrary.

A key indicator of water quality, pH, is regulated by a variety of environmental conditions [21]. The regulation of pH is significantly influenced by dissolved materials, especially minerals and gases. For instance, the water frequently contains significant levels of calcium and magnesium in areas with limestone or dolomite bedrock, resulting in a more alkaline pH [22]. Similar to this, regions with a lot of organic matter that produces CO_2 or with volcanic activity may have more acidic water because dissolved CO_2 causes carbonic acid to develop [23]. Water pH is impacted by pollutants as well. Water's pH can be dramatically lowered by acid rain, which is caused by the atmospheric deposition of pollutants including sulfur dioxide and nitrogen oxides [24]. Heavy metals and other compounds that might change water pH and potentially damage aquatic life as well as degrade water quality are frequently present in industrial effluents [25]. Understanding the variables that affect water pH and keeping an eye on them are therefore essential for maintaining the highest possible level of water quality. This intricacy highlights the necessity for thorough water evaluation methods that can take these various factors into account.

2.3 DO Measurement

Determining the amount of dissolved oxygen in a river is crucial since aquatic life there depends on it and needs at least 4 ppm of DO to survive [8, 18, 26]. DO tests are carried out to confirm this. The Digital Dissolve Oxygen Meter is used to measure the DO test at the location of the sample collection. When measuring DO on-site, the results are correct, but if DO is measured later, the results will alter due to changes in the atmosphere and temperature [10, 14].

2.4 BOD Measurement

The BOD is a measurement of how much total dissolved oxygen is used by microorganisms during the biodegradation of organic materials, such as sewage or food particles. Rural ponds may have an excessive amount of phosphate, cow waste, and domestic sewage from non-point sources, which all contribute to a high organic load and a higher BOD level [27].

2.5 COD Measurement

Chemical oxygen demand, which is typically combined with biological oxygen demand, is a measure of the amount of organics in water (BOD). The COD is a measurement of the amount of organic matter in a sample that can be oxidized by a potent chemical oxidant [28–31]. The COD is frequently employed as a gauge of the organic and inorganic materials' susceptibility to oxidation in water bodies, industrial plant effluents, and sewage discharges [32]. As a result, the COD is an accurate indicator for determining the degree of water pollution (Fig. 3). As organic and inorganic matter concentrations rise, so does the COD of water [33, 34]. According to Garg [35], the COD concentration in Ramsagar reservoir ranged from 3.600 to 17.400 mg/l.

2.6 Carbon Dioxide Measurement

Unrestricted carbon dioxide, an exceedingly soluble gas in water generated by animal respiration, serves as nature's primary source of carbon. It can exist in water as dissolved or bound bicarbonate or carbonates in coral reef habitats, the earth's crust, and limestone. The pH of any system, especially those with inadequate buffering, is reduced when carbon dioxide is dissolved in water. This pH decrease can be harmful to aquatic life [37].

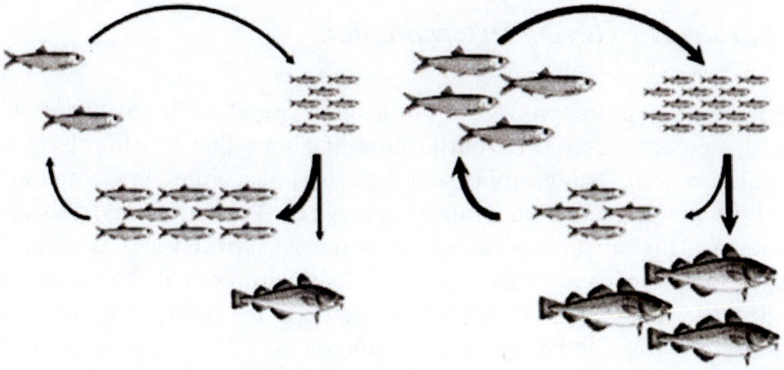

Fig. 3 Community conditions resulting in cod Allee effect: influence of predation on sprat population stages [36]

2.7 *Alkalinity Measurement*

Alkalinity, a measurement of the inclusive concentration of bases in pond water that includes dissolved calcium, magnesium, and other compounds as well as, bicarbonates, hydroxides, borates, phosphates, and carbonates, is the volume of water to engross variations in pH. Alkalinity is primarily increased by lime leaching from cement ponds or calcareous rocks, photosynthesis, sulfate reduction, and denitrification and it is primarily decreased or consumed by sulfide, respiration, and nitrification [38], as well as to a lesser extent by evaporation and decomposing organic matter. However, in case of insufficient alkalinity, even minor bit of acid might source a large pH change.

2.8 *Nitrite-N (NO_2-N) Measurement*

An intermediate in the transformation of NH_4^+ or NH_3 into NO_3, nitrite, is a by-product of oxidized NH_3 or NH_4^+. Nitrification, which is carried out by the system's naturally occurring highly chemo-autotrophic bacteria, aerobic, gram-negative, completes this process. Due to the rapid nature of the dialogue, large nitrite amounts are not frequently seen. The "brown blood illness" is a condition where high amounts of the substance do, however, result in hypoxia brought on by the hemoglobin in fish blood becoming inactive [39]. Boyd [40] asserts that less than 0.3 mg/l of nitrite is the ideal level for aquaculture. The concentration of nitrite at cage stations was greater than the standard permitted value due to the influence from fish waste and more feed, according to research [41–47] that found the concentration of nitrite-N lay in the range of 0.001–0.28 mg/l in cage culture systems. It has been discovered that a higher less dissolved oxygen, more ammonia, and pH can make it more poisonous.

2.9 Nitrate–N (NO_3-N) Determination

Nitrate is created via the process of nitrification, which is the oxidation of NO_2 to NO_3 by aerobic bacteria. De-nitrification of nitrate that aquatic plants cannot directly absorb occurs in anaerobic sediments [48]. According to Garg et al. [35], the ideal nitrate content for aquaculture is between 0.2 and 10 mg/l. Sewage and other pollutants high in nitrates can also contaminate surface water. Drinking water with higher nitrate content is harmful [49]. The Environmental Protection Agency (EPA) has determined that 10.00 mg/l of nitrate is the maximum contamination level (MCL) for water to be used for drinking purpose [50]. The concentration of nitrate in receiving waters is significantly impacted by runoff from farming operations and waste disposal sites. The usage of fertilizers on farms may have contributed to receiving streams' elevated nitrate levels through leaching and surface runoff during heavy rains [34].

2.10 Ammonia (NH_3) Determination

Ammonia is a consequence of the protein metabolism that fish harvest, as well as the breakdown of bacterial of biological waste like dead planktons, food scraps, and manure. Both ionized or unionized and ammonia are referred to as "total ammonia," yet only the unionized form (NH_3) is extremely toxic [36].

2.11 TSS Measurement

Total suspended solids, also known as non-filterable solids, are those solids that are retained by a filter with 1-micron holes. By running a known volume of sewage sample through a glass fiber filter equipment and weighing the dry residue that remains, their amount may be calculated [51].

2.12 Total Dissolved Solids (TDS) Measurement

Total dissolved solids, also known as filterable solids, are those solids that are not retained by a filter with 1-micron holes. By running a known amount of sewage sample through a glass fiber filter equipment and weighing the dry residue that remains after boiling the filtrate sewage, their quantity may be calculated. TDS is made up of organic and inorganic materials that are suspended in water [52]. The taste and potability of water can be impacted by high TDS [53]. TDS levels in drinking water should be measured using a conductivity metre and should be between 500 and 1000 mg/l according to WHO. TDS levels can be managed with treatments like reverse osmosis [54].

2.13 Measurement of Turbidity

As soon as you can after the sample is taken, determine the turbidity. Before examination, gently stir all samples to achieve a representative measurement. If sample preservation is not possible, start the analysis right away. If storage is necessary, refrigerate or cool to 4 °C to reduce the amount of microbial degradation of solids. For the best outcomes, test the turbidity right away without changing the pH or temperature of the original sample. A digital turbidity meter is used to measure turbidity. This was calibrated using a standard NTU [51, 55].

2.14 Heavy Metals Measurement

Each sample of acid-digested water weighed 100 cm^3 and was placed in a beaker. The beaker was then placed in a 70 °C oven to condense the water to 50 cm^3. Using an atomic absorption spectrometer, the concentrations of Co, Cu, Cr, CD, Zn, PB, and Ni in each sample of water were measured (AAS: AAS: LABINDIA: AA-7000) Prior to analysis, AAS required phase of acid digestion, which was accomplished by exposing the samples to concentrated HNO_3 [34]. Samples are digested using the methods endorsed by the American Public Health Association (APHA). The suggestions in the instruction manual were used to determine the instrument settings (IO).

2.15 Chlorides Measurement

Chlorides, which are often present as sodium chloride, alter the flavour of water [53]. Both pipes and vegetation may be harmed by presence of excess chlorides [56]. Titration techniques or ion chromatography are frequently used to determine the concentration of chloride [52]. A 250 mg/l limit is advised by the WHO [53]. Chloride levels can be controlled by reducing industrial effluents and recycling [54].

2.16 Biological Indicators

The presence of specific microbes, algae, and other lifeforms can reveal information about the quality of the water. E. coli is frequently tested for because its presence suggests fecal contamination [53]. Eutrophication can result from an algae bloom, which is frequently brought on by too many nutrients (such phosphates) [57]. Chemical and biological testing work together to offer a complete picture of the water quality [4, 6].

2.17 *Index of Water Quality (WQI) Measurement*

The WQI is premeditated in three steps. Each criterion was given a weight (W_i) in the first phase based on its relative significance to the overall quality of water intended for consumption. Nitrate has been given a maximum weight of 5 due to its significant role in determining the water quality [58]. The relative weight is computed using Eq. 1 in the second step.

$$W_i = \frac{w_i}{\sum_{i=1}^{n} w_i} \quad (1)$$

where n is the number of parameters, relative weight is denoted by W_i, and w_i represents the weight of each parameter. The parameter's calculated W_i values are provided in Table 1.

In the third stage, by dividing each parameter's concentration in each water sample by the relevant standard, according to the BIS specifications, and multiplying the result by 100, a quality rating scale (q_i) is obtained for each parameter (Eq. 2)

$$q_i = \left(\frac{C_i - C_{io}}{S_i - C_{io}}\right) \times 100 \quad (2)$$

where S_i is the BIS-recommended standard for Indian drinking water in mg/l for each chemical parameter and C_i is the concentration of each chemical parameter in every single water sample in mg/l. The parameter's optimal value in clear water is C_{io}. With pH, q_i is equal to $(C_i - 7)/(S_i - 7)100$ and C_{io} is equal to 7. For the remaining parameters, the optimal value is 0.

Table 1 Chemical parameters and their relative weight [59]

Chemical indicators	Weight	Indian standard	Relative weight (W_i)
Magnesium	2.00	30.00	0.091
Total hardness	2.00	300.00	0.091
Calcium	2.00	75.00	0.091
TDS	4.00	500.00	0.182
Chloride	3.00	250.00	0.136
pH	4.00	6.50–8.50	0.182
Nitrate	5.00	45.00	0.227

Table 2 Classification of water quality based on WQI value [59]

Water quality index	Quality of water
>300.00	Unsuitable
200.00–300.00	Very poor
100.00–200.00	Poor
50.00–100.00	Good
<50.00	Excellent

Each parameter's sub-index (SI), which is utilized to calculate the WQI in accordance with Eqs. 3 and 4, is first determined in order to calculate the WQI.

$$\mathrm{SI}_i = W_i \times q_i \tag{3}$$

$$\mathrm{WQI} = \sum_{i=1}^{n} \mathrm{SI}_i \tag{4}$$

The *i*th parameter's subindex is SI_i. The estimated WQI values are classified into five categories in Table 2.

The five categories range from "good water" to "unfit for drinking water," according to the computed WQI values (Table 2). The WQI has an inter-seasonal range of 28–267, with 24% of water bodies rated outstanding in the pre- and post-monsoon seasons and 18% rated excellent during the monsoon season. Approximately 62%, 58%, and 60% of water bodies are in good quality during the monsoon, pre-monsoon, and post-monsoon seasons, respectively in pre-monsoon, monsoon, and post-monsoon seasons, respectively. 14, 16, and 12% of water bodies were found to be poor water bodies in the respective seasons. The remaining 6, 2, and 4% of water bodies were very poor water bodies in aforesaid seasons. 16 and 48% of water bodies are in excellent and good water bodies throughout the monsoon and post-monsoon seasons. Limits for the physicochemical properties of lake water in India are shown in Table 3.

3 Advantageous and Limitations of Water Quality Evaluation and Treatment Methods

Numerous evaluation and treatment techniques have been created throughout the years in the quest to monitor and manage water quality. While each of these approaches has its own advantages and drawbacks, it is important to be aware of both in order to choose the approach that will work best in a certain circumstance. The biological, chemical, and physical testing approaches that are frequently employed in water quality evaluation and treatment are examined in this section along with their benefits and limitations [61]. By doing this, we intend to provide a thorough

Table 3 Physical and chemical features of Indian lake water [60]

Sr.	Parameter	BIS specification
1	PH	6.50–8.50
2	Color	5 Hazena max
3	Appearance	Clear
4	Turbidity	5 NTU max
5	Alkalinity	200 mg/L max
6	EC	Lacking mention
7	Chloride	250 mg/L max
8	Florid	1 mg/L max
9	Sulfate	200 mg/L max
10	Phosphate	Lacking mention
11	Ca H	75 mg/L max
12	T H	300 mg/L max
13	TDS	500 mg/L max
14	Mg H	30 mg/L
15	FRC	0.25 mg/L
16	Silica	Lacking mention
17	COD	Lacking mention
18	Hydrazine	Lacking mention
19	DO	Lacking mention
20	BOD	Lacking mention
21	NO_3	50 g/L
22	SO_3	Lacking mention

grasp of these techniques, permitting more informed choices about the management of water quality.

3.1 *Physical Testing Methods: Benefits and Limitations*

Advantages:

1. Physical tests can frequently be performed on-site and offer fast findings.
2. These tests can often be carried out more easily and affordably than chemical or biological ones.

Disadvantages:

1. Limited Information: Physical examinations can't tell you anything about particular chemical or biological pollutants.

2. Subjectivity: Some physical tests, such as smell and color, can be a little subjective and may not give accurate quantitative information.

In order to obtain a thorough understanding of water quality, each of these techniques should preferably be utilized in combination.

3.2 Chemical Testing Methods: Benefits and Limitations

Advantages:

1. Specificity: Chemical analyses can locate particular contaminants, enabling focused responses to contamination.
2. Quantitative Information: These tests offer quantitative information that can be used to determine conformity with rules and specifications.

Disadvantages:

1. Expensive: Testing for a wide range of potential contaminants can be expensive when doing chemical analyses.
2. Point-in-Time Data: These tests offer a snapshot at a particular moment in time, and it's possible that they don't accurately capture changes in water quality over time.

3.3 Biological Testing Methods: Benefits and Limitations

Advantages:

1. Sensitive Indicators: Some species may be exceptionally sensitive to particular pollutants, making them reliable markers of harmful chemicals.
2. Long-term monitoring: Biological testing is helpful for long-term monitoring since it offers a snapshot of water quality across time rather than only at one particular instant.
3. Threats other than chemical pollutants can be found through biological testing: These include threats like temperature fluctuations, habitat damage, and more.

Disadvantages:

1. Requires professional expertise: Accurate identification and interpretation of biological indicators can be time-consuming and require professional expertise.
2. Seasonal variations make it challenging to compare results obtained at various times of the year since biological indicators can alter with the seasons.

3.4 Water Quality Index (WQI) Method: Benefits and Limitations

Advantages:

1. Simpleness: The WQI method's biggest benefit is the straightforward nature of the procedure. It reduces complicated water data to a single, simple number. With the use of this one number, the general public and decision-makers can be properly informed on the water quality.
2. Comprehensive: The WQI approach provides a comprehensive snapshot of the water quality by accounting for a wide range of characteristics. When assessing the effects of particular actions or tracking changes over time, this can be helpful.
3. Comparative Evaluation: WQI enables users to compare the water quality of one body of water to another or to the same body over time. This can assist in locating issue regions or water quality trends.

Disadvantages:

1. Lack of Specificity: Despite offering a helpful overall evaluation, the WQI does not draw attention to particular contaminants or characteristics. Even in the presence of high levels of a particular pollutant pollution, a location may still have a positive WQI if all other factors fall below permissible limits.
2. Equal Weight: WQI anticipates that all parameters will be given equal weight. This can be a drawback because some criteria might be worse for the environment and human health than others.
3. Variation in Standards: It is difficult to apply a global WQI since different countries and regions have different acceptable limitations for the criteria they utilize.

4 Physicochemical Review of Some Water Samples from India

The study of physical–chemical parameters is crucial to obtaining a precise understanding of the water quality, and it allows us to compare the results of various chemical and physical parameter values with reference values (Table 3). Here is a brief summary of some previous research studies. Pulp and paper sector effluents are one of India's most polluting industries, according to Tripathi et al. [62]. Small and large pulp and paper mills use several procedures that produce effluents with drastically varying physical–chemical characteristics due to their disparate production capacities and raw material sources. To diminish the impact of various pollutants on the environment, before being released into the drainage channel, such contaminated effluents must undergo sufficient treatment. Based on the findings presented here, it can be stated that the effluent emitted from both paper industries is extremely contaminated and exceeds the limits set by Indian regulatory agency norms. Furthermore, it is claimed that the pollutants produced at various stages of the paper manufacturing process can be reduced either by replacement of some current bleaching and pulping techniques with ECF (Elemental Chlorine Free bleaching), bio-pulping, TCF (Total Chlorine Free bleaching), bio-bleaching, and ozone bleaching or by treating the effluent with physical, biological, or chemical methods. This monitoring research of pulp and paper mill effluent collected from wood-based and agro-based sources resulted in the creation of a data bank that may be used to successfully anticipate their toxicity and implement appropriate management. Agrawal et al. [63] concluded that the water in the Bihar River can perform as a good habitat, provide educational data, and aid in understanding water features. The pH value suggests that the alkaline water in May have resulted from the high temperature, which suggests that CO_2 is soluble in the water. The Bihar River's Baba Ghat's water quality metrics, including alkalinity, total hardness, chloride ion, pH, BOD, COD are all well within the acceptable range, according to the analysis. Consequently, appropriate methods should be developed. According to research by Kaur et al. [64], the majority of freshwater bodies worldwide are becoming contaminated by human waste, industrial waste, sewage, agricultural waste, and religious practices like idol immersion. The Central Pollution Control Board [65] has created a thorough set of regulations regarding the act of immersing idols in oceans, lakes, and rivers (Guidelines for Idol Immersion, CPCB 2006). These regulations outline and define the function of state pollution control boards in evaluating the water quality of aquatic bodies and categorizing them according to specific physiochemical parameters. If these recommendations are followed and put into practices, the quality of the river's water after idol immersion could significantly improve. Khan et al. [66] assert that the most crucial metric for determining the water quality is DO, which was discovered to be Nil at seven places despite being necessary for all aquatic creatures' metabolism. Altaf and Saltanat [67] came to the conclusion that water temperature and water chemistry are the primary influences on phytoplankton dispersal, chloride, transparency, dissolved oxygen, alkalinity, TDS, and CO_2. Hardness and Conductivity, on the other hand, have less of an impact on the distribution of phytoplankton groups. Rani Lake water displays low high COD, BOD, DO, hardness, turbidity, alkalinity, chloride, TDS, nitrate, and phosphate during 2008 and 2009, according to evaluations done by Mishra et al. [68]. These parameters' values were discovered to be over the allowable limit (IS: 10500). A higher pH number implies that the water is slightly alkaline. The results unambiguously show that sewage discharge and other anthropogenic activities have caused this lake to become contaminated and entropic in character. The amount of oxygen utilized during aerobic processes of organic material degradation brought on by microorganisms is measured by biological oxygen demand. Therefore, the BOD provides details on the proportion of an organic substance in a sample of water that can be converted biologically. As a result, it is necessary to take into account how susceptible certain materials are to oxygen oxidation. The pre- during and post-immersion activities yielded the ranges of 1.4–35, 4–55, and 3.3–38 mg/L. At Site-7, the immersion period produced the highest BOD value (Majnu Ka Tila). Analysis was also done on the COD, which is frequently used to estimate indirectly how many organic chemicals are present in water. COD is a helpful indicator of water quality because the majority of its applications estimate the volume of organic contaminants present in wastewater or surface water (such as lakes and rivers). As indicated by a number of researchers [69–73], the value of COD in conjugation with BOD is beneficial in determining the hazardous circumstances and existence of physiologically resistant organic compounds.

5 Conclusions and Recommendations

The water quality of rivers and lakes has declined as a result of all the aforementioned effects. It provides a hint for municipal authorities to create effective management plans. The main factors influencing temperature variation are atmospheric temperature and weather. The industrial effluents that were released may have contributed to the pH decrease in Indian water bodies. This outcome is consistent with prior scientists' reports. The decrease in water volume due to rapid evaporation at high temperatures, high loadings of organic compounds, chlorides, detergent, and other pollutants are all responsible for the increase in hardness. Low levels of DO as a result of wastewater discharge from enterprises that were discharging some biological wastes that demanded a lot of oxygen. An important factor that is frequently used to estimate the pollutant load of wastewater is biological oxygen demand. The purpose of the BOD test is to quantify the quantity of carbonaceous matter that is biochemically oxidized. Values of BOD were caused by a faster rate of organic matter decomposition with turbidity, at a higher temperature and slower water current. A test called chemical oxygen demand is used to gauge how polluting household and industrial waste is. This provides important information regarding the possibility for pollution from home sewage and industrial effluents. Many aquatic zones in India have high COD value, which shows that industrial effluents released by industrial units are mostly to blame for the pollution there. Therefore, it is crucial to check the water's purity using the several criteria listed above before utilizing it.

References

1. Shukla, B.K., Bhowmik, A.R., Raj, R.B., Sharma, P.K.: Physico-chemical parameters and status of ground water pollution in Jalandhar-Phagwara region. J. Green Eng. **9**(2), 212–223 (2019)
2. Shenoy, A., Shukla, B.K., Bansal, V.: Sustainable design of textile industry effluent treatment plant with constructed wetland. Mater. Today: Proc. **61**(02), 537–542 (2022)
3. Shenoy, A., Bansal, V., Shukla, B.K.: Treatability of effluent from small scale dye shop using water hyacinth. Mater. Today: Proc. **61**(02), 579–586 (2022)
4. Shukla, B.K., Bashir, M., Sharma, P.K.: An analytical investigation of surface water quality and pollution status in Srinagar, Jammu and Kashmir, India. J. Green Eng. **11**, 952–962 (2021)
5. Madhuri, T.U., Subhashini, B.: A study on groundwater quality in residential colonies of Visakhapatnam. Ecobiol. Polluted Waters, 102 (2006)

6. Shukla, B.K., Gupta, A., Sharma, P.K., Bhowmik, A.R.: Pollution status and water quality assessment in pre-monsoon season: a case study of rural villages in Allahabad district, Uttar Pradesh, India. Mater. Today: Proc. **32**, 824–830 (2020)
7. Periyasamy, S., Temesgen, T.: Application of biochar for wastewater treatment. In: Biochar and its Application in Bioremediation, pp. 363–380. Springer, Singapore (2021)
8. Shukla, B.K., Goel, A.: Study on oxygen transfer by solid jet aerator with multiple openings. Eng. Sci. Technol. Int. J. **21**(2), 255–260 (2018)
9. Patil, P.N., Sawant, D.V., Deshmukh, R.N.: Physico-chemical parameters for testing of water-a review. Int. J. Environ. Sci. **3**(3), 1194 (2012)
10. Shukla, B.K., Khan, A., Saikiran, G., Sriram, M.: Comparative study on effect of variation in opening shape on oxygenation performance of surface jet aerators used in water and wastewater treatment. J. Green Eng. **9**(3), 427–440 (2019)
11. Adebowale, K.O., Agunbiade, F.O., Olu-Owolabi, B.I.: Impacts of natural and anthropogenic multiple sources of pollution on the environmental conditions of Ondo State coastal water, Nigeria. Electron. J. Environ., Agric. Food Chem. **7**(4), 2797–2881 (2008)
12. Edema, M.O., Omemu, A.M., Fapetu, O.M.: Microbiology and Physicochemical Analysis of different sources of drinking water in Abeokuta, Nigeria. Niger. J. Microbiol. **15**(1), 57–61 (2001)
13. Rim-Rukeh, A., Ikhifa, O.G., Okokoyo, A.P.: Effects of agricultural activities on the water quality of Orogodo River, Agbor Nigeria. J. Appl. Sci. Res. **2**(5), 256–259 (2006)
14. Shukla, B.K., Goel, A., Sharma, P.K.: An experimental study on oxygenation performance of rectangular shaped solid jet aerator having rounded edges. J. Green Eng. **10**, 2728–2745 (2020)
15. Dhakyanaika, K., Kumara, P.: Effects of pollution in River Krishni on hand pump water quality. J. Eng. Sci. Technol. Rev. **3**(1), 14–22 (2010)
16. Rajagopal, T., Thangamani, A., Sevarkodiyone, S.P., Sekar, M., Archunan, G.: Zooplankton diversity and physico-chemical conditions in three perennial ponds of Virudhunagar district, Tamil Nadu. J. Environ. Biol. **31**(3), 265–272 (2010)
17. Eruola, A.O., Ufoegbune, G.C., Eruola, A.O., Awomeso, J.A., Adeofun, C.O., Idowu, O.A.: An assessment of the effect of industrial pollution on Ibese River, Lagos, Nigeria. Afr. J. Environ. Sci. Technol. **5**(8), 608–661 (2011)
18. Shukla, B.K., Rajesh Kumar, V., Goel, A.: Experimental studies on the effect of variation in jet length on oxygenation performance of elliptical shaped solid jet aerator. J. Adv. Res. Dyn. Control Syst. **10**(08-Special issue), 1037–1044 (2018)
19. Koirala, S., Jung, M., Reichstein, M., de Graaf, I.E., Camps-Valls, G., Ichii, K., Papale, D., Ráduly, B., Schwalm, C.R., Tramontana, G., Carvalhais, N.: Global distribution of groundwater-vegetation spatial covariation. Geophys. Res. Lett. **44**(9), 4134–4142 (2017)
20. Agarwal, A., Saxena, M.: Assessment of pollution by physicochemical water parameters using regression analysis: a case study of Gagan river at Moradabad-India. Adv. Appl. Sci. Res. **2**(2), 185–189 (2011)
21. Srinivasamoorthy, K., Chidambaram, S., Prasanna, M.V., Vasanthavihar, M., Peter, J., Anandhan, P.: Identification of major sources controlling groundwater chemistry from a hard rock terrain—a case study from Mettur taluk, Salem district, Tamil Nadu, India. J. Earth Syst. Sci. **117**, 49–58 (2008)
22. Schulte, P., Van Geldern, R., Freitag, H., Karim, A., Négrel, P., Petelet-Giraud, E., Probst, A., Probst, J.L., Telmer, K., Veizer, J., Barth, J.A.: Applications of stable water and carbon isotopes in watershed research: weathering, carbon cycling, and water balances. Earth Sci. Rev. **109**(1–2), 20–31 (2011)
23. Smedley, P.L., Kinniburgh, D.G.: A review of the source, behaviour and distribution of arsenic in natural waters. Appl. Geochem. **17**(5), 517–568 (2002)
24. Likens, G.E., Driscoll, C.T., Buso, D.C.: Long-term effects of acid rain: response and recovery of a forest ecosystem. Science **272**(5259), 244–246 (1996)
25. Zheng, N., Liu, J., Wang, Q., Liang, Z.: Health risk assessment of heavy metal exposure to street dust in the zinc smelting district, Northeast of China. Sci. Total Environ. **408**(4), 726–733 (2010)

26. Gupta, S., Bhatnagar, M., Jain, R.: Physico-chemical characteristics and analysis of Fe and Zn in tubewell water and sewage water of Bikaner City. Asian J. Chem. **15**(2), 727 (2003)
27. Dan'Azumi, S., Bichi, M.H.: Industrial pollution and heavy metals profile of Challawa River in Kano, Nigeria. J. Appl. Sci. Environ. Sanit.Ion **5**(1), 23–29 (2010)
28. Sanap, R.R., Mohite, A.K., Pingle, S.D., Gunale, V.R.: Evaluation of water qualities of Godawari River with reference to physico-chemical parameters, Dist. Nasik (MS), India. Pollut. Res. **25**(4), 775 (2006)
29. Pande, K.S., Sharma, S.D.: Natural purification capacity for Ramganga River at Moradabad (UP). Pollut. Res. **17**, 409–415 (1998)
30. Murhekar, M.V., Murhekar, K.M., Das, D., Arankalle, V.A., Sehgal, S.C.: Prevalence of hepatitis B infection among the primitive tribes of Andaman & Nicobar Islands. Indian J. Med. Res. **111**, 199–203 (2000)
31. American Public Health Association: APHA Standard Methods for the Examination of Water and Wastewater. Standard Methods for the Examination of Water & Wastewater. American Public Health Association, Washington, DC (2005)
32. Chapman, D.: Water Quality Assessments: A Guide to the Use of Biota, Sediments and Water in Environmental Monitoring. CRC Press (2021)
33. Boyd, C.E.: Water Quality in Warmwater Fish Ponds. Craftmaster Printers. Inc., Opelika, Alabama (1979)
34. Sharma, P.K., Ayub, S., Shukla, B.K.: Cost and feasibility analysis of chromium removal from water using Agro and horticultural wastes as adsorbents. In: 3rd International Conference on Innovative Technologies for Clean and Sustainable Development: ITCSD 2020 3, pp. 449–463. Springer International Publishing (2021)
35. Garg, R.K., Rao, R.J., Uchchariya, D., Shukla, G., Saksena, D.N.: Seasonal variations in water quality and major threats to Ramsagar reservoir, India. Afr. J. Environ. Sci. Technol. **4**(2), 61–76 (2010)
36. Van Leeuwen, A., De Roos, A.M., Persson, L.: How cod shapes its world. J. Sea Res. **60**(1–2), 89–104 (2008)
37. Verma, D.K., Satyaveer, N.K.M., Kumar, P., Jayaswa, R.: Important water quality parameters in aquaculture: an overview. Agric. Environ. **3**(3), 24–29 (2022)
38. Butler, J.N.: Aquatic chemistry: an introduction emphasizing chemical equilibria in natural waters (Stumm, Werner) (1971)
39. Lawson, T.B. (Ed.): Fundamentals of Aquacultural Engineering. Springer Science and Business Media (1994)
40. Diana, J.S., Szyper, J.P., Batterson, T.R., Boyd, C.E., Piedrahita, R.H.: Water quality in ponds. Dyn. Pond Aquac., 53–71 (2017)
41. Siti-Zahrah, A., Padilah, B., Azila, A., Rimatulhana, R., Shahidan, H.: Multiple streptococcal species infection in cage-cultured red tilapia but showing similar clinical signs. In: Diseases in Asian Aquaculture VI. Manila: Fish Health Section, Asian Fisheries Society, 2008, pp. 313–320 (2008)
42. Eglal, A.O., Nour, A.M., Essa, M.A., Zaki, M.A., Mabrouk, H.A.: Technical and economical evaluation of small-scale fish cage culture for youth in the River Nile for Egypt: 1. Effect of stocking density of mono-sex Nile Tilapia (Oreochromis niloticus) fingerlings. In: Cage Aquaculture in Asia: Proceeding of the International Symposium on Cage Aquaculture in Asia, pp. 107–114 (2009)
43. Mondal, M.N., Shahin, J., Wahab, M.A., Asaduzzaman, M., Yang, Y.: Comparison between cage and pond production of Thai Climbing Perch (Anabas testudineus) and Tilapia (Oreochromis niloticus) under three management systems. J. Bangladesh Agric. Univ. **8**(2), 313–322 (2010)
44. Jiwyam, W.: Extensive net cage culture of Nile Tilapia (Oreochromis niloticus) fingerlings in nutrient-enriched pond. Our Nat. **10**(1), 61–70 (2012)
45. Nyanti, L., Hii, K.M., Sow, A., Norhadi, I., Ling, T.Y.: Impacts of aquaculture at different depths and distances from cage culture sites in Batang Ai Hydroelectric Dam Reservoir, Sarawak, Malaysia. World Appl. Sci. J. **19**(4), 451–456 (2012)

46. Gorlach-Lira, K., Pacheco, C., Carvalho, L.C.T., Melo Júnior, H.N., Crispim, M.C.: The influence of fish culture in floating net cages on microbial indicators of water quality. Braz. J. Biol. **73**, 457–463 (2013)
47. Yee, L.T., Paka, D.D., Nyanti, L., Ismail, N., Emang, J.J.: Water quality at Batang Ai hydroelectric reservoir (Sarawak, Malaysia) and implications for aquaculture. Int. J. Appl. **2**(6), 23–30 (2012)
48. Furnas, M.J.: The behavior of nutrients in tropical aquatic ecosystems. In: Pollution in Tropical Aquatic Systems, pp. 29–65. CRC Press (2018)
49. Manickam, N., Bhavan, P.S., Santhanam, P., Bhuvaneswari, R., Muralisankar, T., Srinivasan, V., Asaikkutti, A., Rajkumar, G., Udayasuriyan, R., Karthik, M.: Impact of seasonal changes in zooplankton biodiversity in Ukkadam Lake, Coimbatore, Tamil Nadu, India, and potential future implications of climate change. J. Basic Appl. Zool. **79**(1), 1–10 (2018)
50. Self, J.R., Waskom, R.M.: Nitrates in Drinking Water. Service in Action, no. 0.517 (2008)
51. Gorde, S.P., Jadhav, M.V.: Assessment of water quality parameters: a review. J. Eng. Res. Appl. **3**(6), 2029–2035 (2013)
52. Rice, E.W., Bridgewater, L., American Public Health Association (eds.): Standard Methods for the Examination of Water and Wastewater, vol. 10. American Public Health Association, Washington, DC (2012)
53. Gordon, B., Ahmed, F., Chorus, I., Cotruvo, J., Cunliffe, D., de Roda Husman, A.: Guidelines for drinking-water quality: World Health Organization (2011)
54. Xu, P., Drewes, J.E.: Viability of nanofiltration and ultra-low pressure reverse osmosis membranes for multi-beneficial use of methane produced water. Sep. Purif. Technol. **52**(1), 67–76 (2006)
55. Sawant, R., Chavan, N.: Water quality status of Mahagaon reservoir from Gadhinglaj Tahsil from Maharashtra. Int. J. Sci. Environ. Technol. **2**(6), 1196–1204 (2013)
56. Yahya, S.M., Rahman, A., Chughtai, M.O.: Study on environmental impact of chemical pollutants & interpretation of characteristics using statistical technique in water quality of Keenjhar Lake located in Sindh Pakistan. J. Mater. Environ. Sci. **7**(2), 648–659 (2016)
57. Conley, D.J., Paerl, H.W., Howarth, R.W., Boesch, D.F., Seitzinger, S.P., Havens, K.E., Lancelot, C., Likens, G.E.: Controlling eutrophication: nitrogen and phosphorus. Science **323**(5917), 1014–1015 (2009)
58. Ramakrishnaiah, C.R., Sadashivaiah, C., Ranganna, G.: Assessment of water quality index for the groundwater in Tumkur Taluk, Karnataka State, India. E-J. Chem. **6**(2), 523–530 (2009)
59. Rao, G.S., Nageswararao, G.: Assessment of ground water quality using water quality index. Arch. Environ. Sci. **7**(1), 1–5 (2013)
60. Manjare, S.A., Vhanalakar, S.A., Muley, D.V.: Analysis of water quality using physicochemical parameters Tamdalge tank in Kolhapur district, Maharashtra. Int. J. Adv. Biotechnol. Res. **1**(2), 115–119 (2010)
61. Bhateria, R., Jain, D.: Water quality assessment of lake water: a review. Sustain. Water Resour. Manag. **2**, 161–173 (2016)
62. Tripathi, P., Kumar, V., Joshi, G., Singh, S.P., Panwar, S., Naithani, S., Nautiyal, R.: A comparative study on physico-chemical properties of pulp and paper mill effluent. Int. J. Eng. Res. Appl. **3**(6), 811–818 (2013)
63. Agrawal, N., Choubey, P.M., Pandey, J.P.: Water quality assessment of Baba Ghat of Bihar River Rewa (MP) India. Int. J. Sci. Res. Publ. **4**(10), 306–308 (2014)
64. Kaur, B.J., George, M.P., Mishra, S.: Water quality assessment of river Yamuna in Delhi stretch during Idol immersion. Int. J. Environ. Sci. **3**(6), 2122 (2013)
65. Bhawan, P., Nagar, E.A.: Central Pollution Control Board (2020)
66. Khan, A.A., Shammi, Q.J., Hussain, S.D., Gulam, N.N.: Seasonal variations in physico-chemical parameters in upper lake of Bhopal (MP). Int. J. Appl. Univers. Res. **2**(2), 1–7 (2015)
67. Altaf, H.G., Saltanat, P.: Effect of physico-chemical conditions on the structure and composition of the phytoplankton community in Wular Lake at Lankrishipora, Kashmir. Int. J. Biodivers. Conserv. **6**(1), 71–84 (2014)

68. Mishra, R., Prajapati, R.K., Dwivedi, V.K., Mishra, A.: Water quality assessment of Rani Lake of Rewa (MP), India. GERF Bull. Biosci. **2**(2), 11–17 (2011)
69. Rajurkar, N.S., Nongbri, B., Patwardhan, A.M.: Physico-chemical and microbial analysis of Umiam (Barapani) lake water. Indian J. Environ. Prot. **23**, 633–639 (2003)
70. Shukla, B.K., Sharma, P.K., Goel, A.: Study on oxygenation performance of solid jet aerator having circular opening corresponding to variable jet length and flow area. In: Journal of Physics: Conference Series, vol. 1531, no. 1, p. 012117. IOP Publishing (2020)
71. Gupta, A.K., Kavita, M., Pramod, K., Chandrasekhar, S., Sumita, S.: Impact of religious activities on the water characteristics of prominent ponds at Varanasi (UP), India. Plant Arch. **11**(1), 297–300 (2011)
72. Singh, S., Singh, V.P., Garg, P., Shukla, B.K.: Physico-chemical analysis of groundwater in Noida-Ghaziabad region. In: IOP Conference Series: Earth and Environmental Science, vol. 1110, no. 1, p. 012028. IOP Publishing (2023)
73. Imtiyaz, I., Krishnakant, Shukla, B. K., Varadharajan, S., Bharti, G.: A comprehensive review of mass transfer phenomenon in gas-liquid phase flow during aeration in wastewater treatment. In: Proceedings of Indian Geotechnical and Geoenvironmental Engineering Conference (IGGEC) 2021, vol. 2, pp. 127–135. Springer Nature Singapore, Singapore (2022)

Surface Water Quality Modelling Using Water Quality Index (WQI) and Geographic Information System (GIS) on the Mahanadi Basin, Odisha

Abhijeet Das

Abstract The Mahanadi River basin in Odisha relies heavily on surface water for its drinking water supply. Surface water quality due to geogenic factors, aggravated by anthropogenic activities, is a significant threat to human wellbeing and agricultural practices. Consequently, surface water quality profiling has received remarkable academic attention in recent decades that produces an ample amount of research results. This work makes an effort to comprehend the acceptability of water for human consumption by utilising the effectiveness of Canadian council of ministers of the environment water quality index (CCME WQI) and multi-criteria decision-making (MCDM) approach like Weighted aggregated sum-product assessment (WASPAS). The investigation was conducted in two stages. Surveying the study area and collecting and interpreting water sample were part of the first phase. The observed water quality data sets were subjected to the second phase's application of CCME WQI and WASPAS. 19 sampling sites were used in the research region during the 2022–2023 sampling campaign, which was based on physico-chemical criteria. Spatial maps has been prepared using Arc GIS, on the basis of the selective chosen quality parameters to decipher the various quality classes at each hydro-station for drinking purpose. However, this study conducts a comprehensive systematic literature review to summarize and structure the existing literature and to identify current research trends and hotspots. Reported results revealed that while comparing with WHO guidelines, total coliform (TC) and total Kjeldahl nitrogen (TKN) found to be higher than the permissible, in most of the locations. The CCME WQI in the study ranges from 34.61 to 85.58. It indicates good to poor water. According to percentage distribution, it shows that 73.68% indicates 'good water', 15.79% of samples is in 'fair category' and 10.53% falls under 'poor' water category. The interpolated map displays that 'ST-8 and ST-9' are located in a poor region and require treatment before consumption. Poor zone delivers about anthropogenic inputs such as domestic sewage and agricultural runoff. Employing the WASPAS approach, it indicates areas where cumulative variables, such as sewage discharge, a falling water

A. Das (✉)
Research Scholar, Department of Civil Engineering, C.V. Raman Global University (CGU), Bhubaneswar, Odisha 752054, India
e-mail: das.abhijeetlaltu1999@gmail.com; 2101070005@cgu-odisha.ac.in

S. Saxena et al. (eds.), *Emerging Materials and Technologies in Water Remediation and Sensing*, Lecture Notes in Civil Engineering 439,
https://doi.org/10.1007/978-981-99-6762-9_2

table, dilution and surface runoff, which cause water quality variability in a water body throughout a monitoring period, have had the greatest impact. Three sampling sites (ST-8 [Cuttack D/s], 9 [Paradeep] and 19 [Choudwar D/s]) were identified by the results as having had the most fluctuation and being the most polluted monitoring sites. In order to identify sampling stations for routine monitoring reasons, this study introduces a novel methodology by combining both WQI and WASPAS. According to the current research using both approaches, the majority of the region's surface water is of acceptable quality, with a few exceptions at a few locations, but it still needs to be treated before consumption in order to avoid the dangers of geogenic and anthropogenic contamination. The findings of this innovative work provides insights and understanding on surface water pollution regimes and minimize uncertain causes because of the high spatial distribution of coliform and nitrogen occurrence, and will also be helpful to policy makers for better planning, investments, and management to supply potable water in the area.

Graphical Abstract

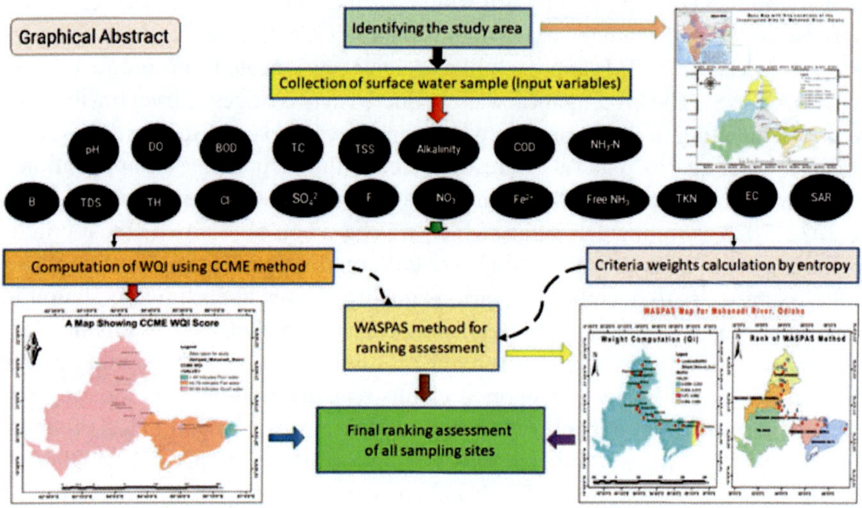

Keywords Mahanadi · CCME · WASPAS · Agricultural · Dilution · Geogenic · Anthropogenic

1 Introduction

The physical inadequacy of the resource is only one factor contributing to the global water shortage; another is the gradual decline in water quality occurring in many nations, which lowers the amount of usable water [1]. Water quality of rivers and their catchments is impacted by a variety of natural and human-made factors [2]. So, an essential tool for preserving public health and water resources is water quality monitoring and assessment [3]. However, pollution renders water unfit for human consumption, and it is difficult and expensive to treat to a level that is adequate for consumption. On the other hand, the quality of drinking water in developing nations is generally poor, especially in rural areas, and is a primary cause of numerous water-borne illnesses. Urbanisation has a significant impact on the hydrologic regime of catchments and the quality of water discharge [4]. Water resources are severely impacted by two key variables, such as industrialization and urbanisation, which also increase the risk of contamination [5]. Most enterprises along the riverside release their wastewater into the neighbouring bodies of water with little to no treatment. Additionally, the primary danger to the quality of surface water is environmental pollution brought on by commercial and residential activity [6]. Farmlands frequently have to use low-quality water due to the constantly rising demand for water [7]. However, the continuing use of water can impair farming productivity and may have negative effects on farmers, product consumers, and the environment [8]. Additionally, irrigated agriculture, industrial growth and urbanisation are all increasing quickly in the study area, especially during the past ten years, as a result of government development policies and strategies [9]. However, the development plans that are predicted to have an impact on drinking water and irrigation agriculture are out of proportion to waste collection and treatment infrastructure [10]. In light of this, it is crucial to evaluate and keep track of the water quality of the accessible resources, which are mostly used for drinking [11]. Numerous studies have been carried out to track and assess the quality of drinking water using specific physical and chemical characteristics as well as a wide range of irrigation water quality indices (WQIs). Hence, WQIs may be evaluated using a straightforward mathematical process that reduces a vast array of water characteristics to a single number that represents the sum of all water quality values [12]. Due to its single value and simplicity, the WQI technique is preferred by a vast number of nations when determining the overall river status [13]. Kong et al. [14] used 10 water quality measures to introduce this methodology for the first time. A composite WQI was created by the CCME (Canadian Council of Ministers of the Environment), and it was employed in this investigation as the water quality integral indicator. It is focused on a variety of water uses, including drinking, recreation, agriculture, animals and sea life [15]. It is a standardised method of evaluating the quality of the water that was created by Canadian authorities [16, 17]. It has progressed from evaluating each individual indicator of water quality to more intricate methods that enable a broader comprehension of the health of the water body and the connection between its natural reaction and human-caused effect [18]. Prioritising decisions to apply policies is a crucial component of monitoring programmes. Multi-criteria decision-making (MCDM) techniques are frequently taken into account in such circumstances. The information obtained by the monitoring system must match the information requirements for water quality, and this gap must be filled. However, because there are so many random events in our environment, Shannon entropy has recently found utility in a number of domains, including environmental engineering and hydraulic engineering. In light of Shannon entropy, a number of techniques have been developed, such as the creation of entropy-based quality indices to identify the shortfalls of traditional WQIs, the alteration of conflicts among several WQIs by entropy-designed MCDMs, and optimising internet monitoring systems to expand on the information learned through monitoring programmes [19]. Weighted aggregated sum-product assessment (WASPAS), an MCDM technique, could be used to identify the worth and utility of different sectors and establish a priority list for implementing them [20]. This method allows the evaluation of a discrete collection of options based on a set of decision criteria to be studied. However, the geographical information system (GIS) technique has been widely employed recently to monitor and assess surface water. Surface water spatial data can be interpreted and analysed using the inverted distance weighted (IDW) interpolation method in conjunction with GIS technology. It is sometimes referred to as a time-saving technique and is used to determine pollution sources, trends and correlations [21]. Surface water serves as the main supplier of water for drinking, in the current study region. However, only a little amount of study has been done on assessing the water quality in the Mahanadi River basin (MRB), in Odisha. There has not been a single study that integrated the CCME WQI, GIS and WASPAS to evaluate the water quality in this area. As a result, there is a research gap, and further debate is required to comprehend the scope and reasons behind the decline of water quality. The main objective of the current study is to identify the most effective framework for estimating water quality in order to provide accurate and effective knowledge at the lowest feasible cost.

2 Materials and Methods

2.1 Description of Study Area

The Mahanadi river system is the biggest river in Odisha and the third-largest river system in the entire Indian subcontinent. About 4% of the country's total geographic area, or 141,600 Km^2, is drained by rivers. Additionally, it can discharge 44,740 m^3/sec at its peak. This river gets effluents from a variety of urban and industrial areas along its course, including Sambalpur, Bauda, Cuttack, Choudwar, Jagatpur and Paradeep in Odisha. At Sambalpur, Cuttack and Paradeep, where the growth of industries and sewer discharges is considerable, human influences are noticeable [2]. From latitudes 19°21′N and 23°35′N and longitudes 80°30′E and 86°50′E will be where the basin is located. It begins in Chhattisgarh's Dhamtari area at a height of around 442 m above mean sea level and flows into the Bay of Bengal. Its length is around 851 kms. Despite extensive research on the impact of various urban and industrial effluents on the Mahanadi river-estuarine system's water quality, the study only looked at a small portion of the river's course [22]. The temperature in this river typically ranges from 15.8° to 28.7 °C and is subject to a tropical monsoon environment. 1360 mm of rain falls annually, largely between the months of June and September with sporadic torrential downpours from cyclonic storms. The dominant land uses are often agricultural, forestry preserves, mining areas and urban areas. On the other hand, red and yellow soils and mixed red and black soils are the major soil varieties in the river basin. In Fig. 1, the precise locations of the effluent discharge stations are shown.

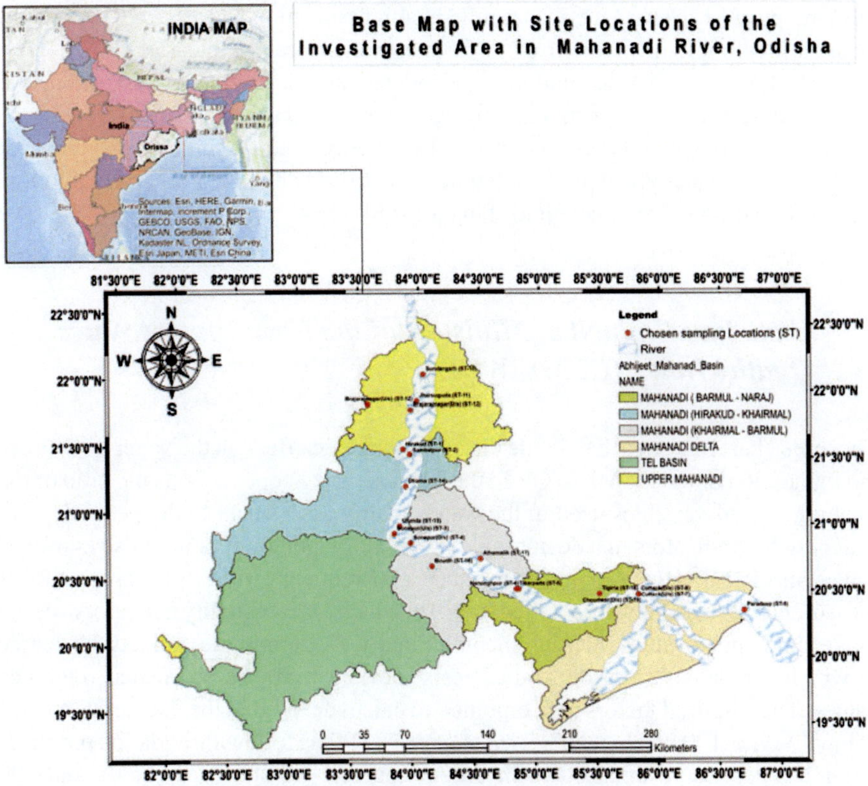

Fig. 1 Map of H_2O sampling sites in the Mahanadi River, Odisha

2.2 Sampling Strategies and Analysis

At 19 locations along the river system's course, water samples were collected for the current study. Over the course of 2022–2023, samples were gathered with the aid of boats. The mean value is employed after considering three sets of data. Twenty physico-chemical parameters such as dissolved oxygen (DO), sulphate (SO_4^{2-}), total alkalinity (TA), free ammonia (free-NH_3), nitrate (NO_3^-), total Kjeldahl nitrogen (TKN), pH, biochemical oxygen demand (BOD), sodium adsorption ratio (SAR), total suspended solids (TSS), total coliform (TC), boron (B^+), total dissolved solids (TDS), chemical oxygen demand (COD), chloride (Cl^-), ammoniacal nitrogen (NH_3-N), fluoride (F^-), electrical conductivity (EC), iron (Fe^{2+}) and total hardness (TH) were selected, which were assessed and compared with WHO (2011) standards for drinking purposes. The American public health association's (APHA 2012) recommended methodologies were used to further evaluate each indicator. CCME WQI and WASPAS were two techniques used to evaluate the overall quality of surface water. A base map of the ongoing study was also created using ArcGIS 10.8, as seen in Fig. 1. Except when otherwise noted, all substances and reagents employed in the analysis were of analytical grade. All dilutions utilised deionized water. By dilution, the stock solutions were made into standard solutions.

2.3 Canadian Council of Ministers of the Environment Water Quality Index (CCME WQI)

The three characteristics that define the anthropogenic effect on the water quality are the foundation of the CCME WQI. In the 1st case, F_1 (Scope) reflects the ratio of the total number of variables used in the water quality assessment to the percentage of water quality indicators that do not satisfy regulatory guideline values (also known as 'failed variables'). However, in 2nd instance, F_2 (frequency) reflects the proportion of measurements (also known as 'failed tests') in which a water quality indicator exceeds the recommended values (measurements). Finally, F_3 (amplitude) reflects the degree to which the values from the 'failed tests' deviate from the associated reference values. The acquired factors are combined to calculate WQI in the manner described below: $WQI = 100 - [(F_1^2 + F_2^2 + F_3^2)^{0.5}/1.732]$. The following equation produces a number between 0 and 100, with 0 denoting the 'worst' water quality and 100 denoting the 'best'. Following this, it categorises water quality into five categories: excellent (95–100), good (80–94), fair (66–69), marginal (45–64) and poor (0–44).

2.4 *Weighted Aggregated Sum-Product Assessment (WASPAS)*

This technique uses a multi-criteria decision-making process to rank the options. However, information entropy is frequently used to describe a numerical indicator of the information, disorder, or uncertainty around the occurrence of a random process, which is used to calculate the weights. The entropy model is created utilising physio-chemical parameters and its schematic diagram of the methodology is demonstrated in Fig. 2. There has been a lot of scientific investigation into the use of information entropy, also referred to as Shannon entropy, in the analysis of hydro-meteorological variables recently. It is one of the reliable utility determining methodologies in 2012. This MCDM strategy combines the weighted product model (WPM) and the weighted sum model (WSM). When there are many collinearities and noise in the input variables and there are more input parameters than output parameters, it is a useful data processing approach. Additionally, by combining the subjective weights assigned by decision-makers with the objective weights produced by the entropy technique, this integration develops to produce new weights for the criterion. Additionally, this combination can produce weights for decision-making criteria that are realistic and improve stability across different weights. This technique's methodology was recommended by Zavadskas et al. [20, 23, 24].

3 Results and Discussions

3.1 *Surface Water Quality Parameters*

In the paragraphs that follow, the various study parameters are explained. The pH, which is neutral, serves as a measure of the concentration of hydrogen ions in water. It stands for the basic or acidic properties of water. pH levels for drinking water must be between 6.5 and 8.5. However, the pH ranges from 7.75 to 7.93 mg/l in the surface water samples that were obtained, demonstrating that is very well within the acceptable range. However, a low value for DO, which is crucial for aquatic life, suggests the oxidation of organic waste and an influx of contaminants [25]. The readings varied from 7.3 to 7.9 mg/l. It exceeds the specified threshold for drinking water (6 mg/l) in almost all places. Hence, adequate amount of DO is available in the study region. Important indicators of organic contamination in water bodies include BOD and COD. BOD is the amount of oxygen required for aerobic bacteria to carry out their metabolic functions during the breakdown of biodegradable organic waste [26]. In the current study, it was detected in the range of 1.1–2.5 mg/l. The average value is found to be low as per WHO standards (5 mg/l) and hence safe for consumption. A measure of the chemical composition of dissolved or suspended organic and inorganic materials in water is called COD. Low value denotes improper treatment of residential wastewater discharges [27]. It typically denotes the presence

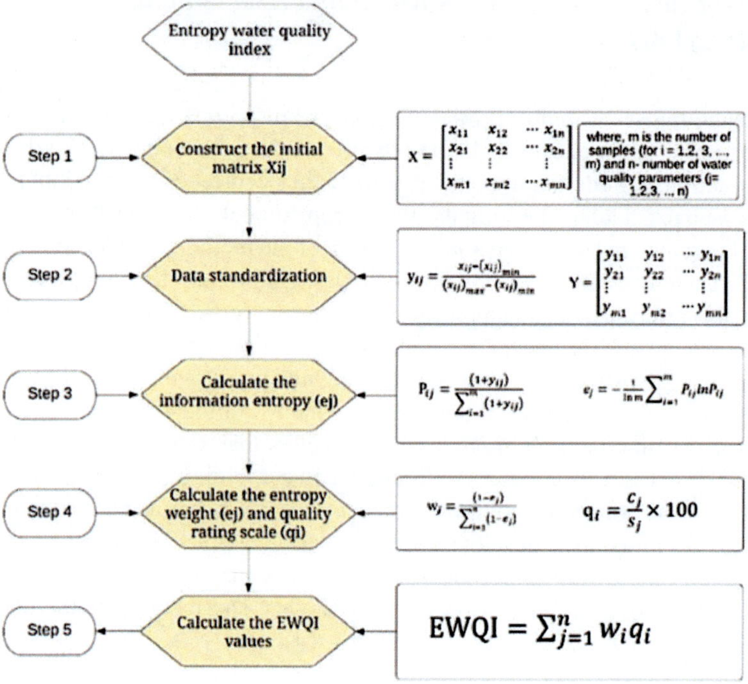

Fig. 2 Framework to calculate entropy weights

of organic materials, primarily nitrites, sulphides and ferrous salts [28]. The values were in the ideal range of 7 to 22 mg/l, which is found to be 30 mg/l. Total coliform (TC) in water comes from sanitary river contamination [29]. Its concentration in the current study ranges from 1250 to 42,900 MPN/100 ml. The chemical industry, sewage, fertilisers and agricultural runoff are all responsible for pollution, as seen by the increased amounts at sites ST-8, 9 and 19. It may have an impact on plant development and bacterial degradation, which together may lower the amount of oxygen in the water [30]. TSS is a by-product of mining operations, primarily during excavation, dust and soil. When exposed soil is not covered by vegetation during the monsoon, the kinetic energy of raindrops and subsequent runoff causes a significant amount of soil to erode and become suspended in the runoff water [31]. Its value ranged between 30 and 75 mg/l, in the ongoing work. Its score was found to be under the acceptable limit (100 mg/l). Water becomes alkaline when CO_3^{2-}, OH^- and HCO_3^- ions are present. Higher alkalinity in water, and vice versa, increases its ability to neutralise acids. It should not be higher than 200 mg/l. For the study area, alkalinity ranges from 71 to 111 mg/l. NH_3-N shows the fresh organic load input into the water system. Municipal sewage discharges are thought to be the main source of organic matter for the river system [32]. The readings ranged between 0.52 and 1.94 mg/l, which is below the threshold (2 mg/l). The amount of sewage and wastewater discharged by two fertiliser plants and big townships is related to the higher value at ST-19. Free-NH_3 confirming that human activity in a specific area is the source of ammonia rather than river discharge [33]. The diluting impact is thought to be particularly pronounced during the monsoon season. The range fluctuates in the current research is found to be 0.025–0.07 mg/l. The acceptable value for drinking is less than 2 mg/l. The water quality at most sites (4–12 mg/l) is higher in organic matter pollution compared to WHO standards (5 mg/l), according to TKN, a very effective predictor of the negative impacts on the water recipients of partially processed or unprocessed waste waters from urban areas, industry and agriculture. Gastric cancer risk and other health risks for newborns and expectant mothers are increased by higher values in drinking water [34]. The parameter 'EC' is used to quantify how well a substance or solution can carry an electric current through water. It directly relates to the dissolved material in a water sample [35]. The preferred EC limit is 2250 µS/cm. In this study, the value varied between 139 and 7781 µS/cm. Given that these sites are close to extensive urbanisation, the high EC at ST-9 shows that sewage has been diluted in surface water. In SAR, soil reacted with sodium might become less permeable. The greater the effect of salt on water infiltration and aeration, the finer the soil texture, the greater the levels of organic matter and so on [33]. The SAR was found to be in the range of 0.42–17 mg/l. Threshold value should be 10 mg/l. Observed higher value at ST-9 (17 mg/l) suggests weathering of silicate minerals (alkali feldspars), which form rocks and/or dissolving of soil salts present there due to evaporation [36]. It has been found that large percentage of rocks and many minerals contain boron. Additionally, it is thought of as a very reactive alkali metal. The surface water samples containing boron (B^+) is in the range of (0.02–0.55 mg/l), which is below the detection limit to 2 mg/l. Inorganic salts and a small number of organic molecules that have been dissolved in water make up the majority of TDS. The flavour of water with a very low TDS is unremarkable. Less than 100 mg/l of TDS is considered appropriate for use in drinking water, according to the WHO. It lies in the range of 83–13,231 mg/l. The taste and hardness of the water can be changed by high TDS at ST-9. The main sources of water pollution at ST-9 include industrial or sewage treatment plants, household runoff, soil contamination caused by leaching and agricultural practises [37]. The calcium and magnesium that are dissolved in the water are measured by TH. Since calcium and magnesium are excellent solvents, water that is travelling through soil and rock dissolves naturally occurring minerals and transports them into the surface water [38]. A hardness of 300 mg/l is the most that can be tolerated. Range of hardness is from 52 to 2196 mg/l. Heart problems and renal stones can both be brought on by elevated ST-9 scores in people. Additionally, it has to do with farming methods and point sources of water pollution from companies and storage facilities. The hazard to human health increases with the concentration of Cl^- in water. The related cation affects the Cl^- ion's taste threshold in water. More than 250 mg/l of chloride should not be present. The range of the quantity in this investigation is 9–4910 mg/l. The rising Cl^- concentrations at ST-9 seem to be caused by geogenic or human-made activities. Sulphate, which refers to dissolved or leached from rocks containing gypsum, iron sulphides and other sulphur-bearing compounds, is a crucial characteristic. Apart from ST-9, the range in the current study is 4.50–378 mg/l, which is significantly below the 200 mg/l acceptable limit. Furthermore,

ST-9 is a centre of industry, adjacent to the factories, and it's possible that rainwater containing SOx from the atmosphere increased the water's sulphate concentration. Due to geological processes, F^- is primarily found in water. Skeletal fluorosis can be brought on by fluoride in high quantities (>1.0 mg/l). The concentration in this study ranges from 0.20 to 0.90 mg/l. $NO_3^-$is a key part of the nitrogen cycle and naturally occurring ions [39]. In excess, it is undesirable because it can lead to blue-baby syndrome in new-borns under 6 months old. In general, if it is present in excess of the allowed level of 45 mg/l, the higher concentration poses health risks. Additional animal wastes, plants and living things are also ammonified within the soil stratum, producing NH_3, which is then oxidised by nitrosomonas and nitrobacter bacteria to produce nitrate [40]. The value in the present study lies between 1.2 and 2.8 mg/l. The weathering of Fe-bearing minerals and rocks is one of the most frequent sources of iron in surface water. Although the iron is present in nature in the reduced Fe^{2+} state, its dissolution raises the concentration of iron in water. In the current study, the level of Fe^{2+} fluctuated between 0.5 and 2.7 mg/l. These numbers fell within the drinking water tolerance (3 mg/l). All the significant parameters that are generated using GIS in the form of geospatial map are represented in Fig. 3a–t.

For the 19 sample sites' water quality examination, the CCME WQI was calculated for the given water quality parameters. The results demonstrate that many characteristics than just one have a significant impact on water quality. The range of the obtained WQI, from good to poor categories, is 34.61–85.58. From 19 water samples, 73.68% of samples are in good water, 15.79% are fair water and the remaining 10.53% indicates poor water. WQI data were interpolated throughout the whole study region in order to construct an index map. In Fig. 4, you can see the completed map. Generally, minimum value of WQI is observed at the upstream side. This is because of no irrigation projects and human activities are less. The map correctly defines ST-8&9 as being in the poor group. This could be brought on by direct effluent discharge, excessive surface water consumption, agricultural consequences and efficient ion leaching. In addition, higher value noticed at some locations is because of high TKN and TC in surface water.

Additionally, MCDM tools like WASPAS were utilised to analyse variances in a sizable complex data set. It enables the retrieval of secret information (rank) from a dataset about the possible impacts of the environment on water quality. This approach is used for 20 samples from 19 sites in this investigation. It was carried out using a matrix of physico-chemical data. The analysis was carried out using MATLAB programme. In Table 1, the WASPAS methodology's overall rating of the sampling locations is displayed. These places had overall WASPAS rankings of 1, 2 and 3, respectively, designating them as the most polluted sites. The identification of the relatively more polluted site (rank 1) is found to be ST-9 (0.909), followed by (rank 2), i.e. ST-8 (0.076) and finally (rank 3), i.e. ST-19 (0.063). However, the WASPAS method gave the sampling sites a general score and worked well for ranking the sites when taking into account the criteria for pollution. The spatially created map based on location coordinates and quality criteria is shown in Fig. 5. The different causes include soil leaching and anthropogenic influences such as the admixture of chemicals and animal wastes during irrigation and the presence of wastewater in water

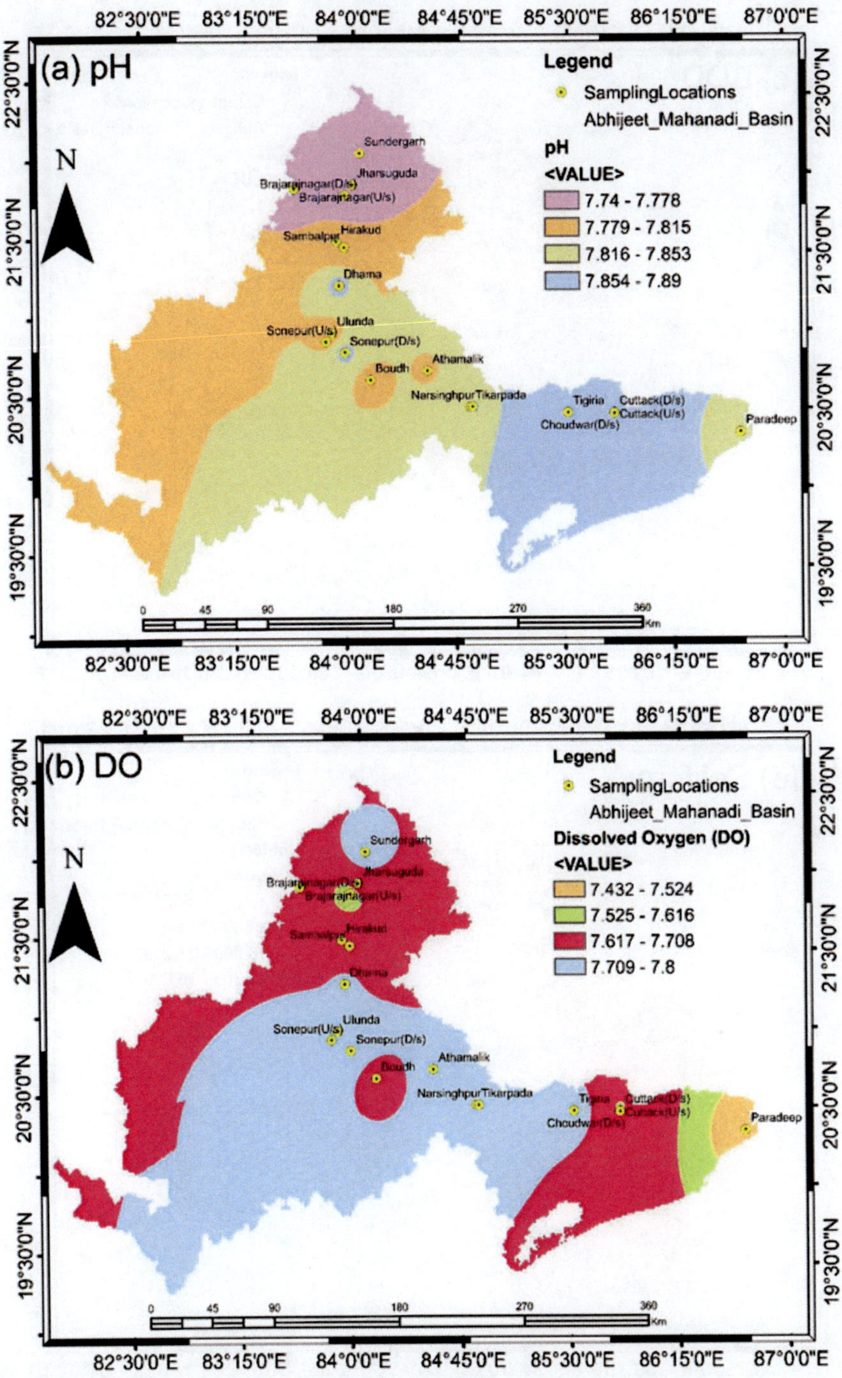

Fig. 3 Spatial distribution map of **a** pH, **b** DO, **c** BOD, **d** TC, **e** TSS, **f** TA, **g** COD, **h** NH_3-N, **i** free NH_3, **j** TKN, **k** EC, **l** SAR, **m** B, **n** TDS, **o** TH, **p** Cl^-, **q** $SO_4{}^{2-}$, **r** F^-, **s** $NO_3{}^-$ and **t** Fe^{2+}

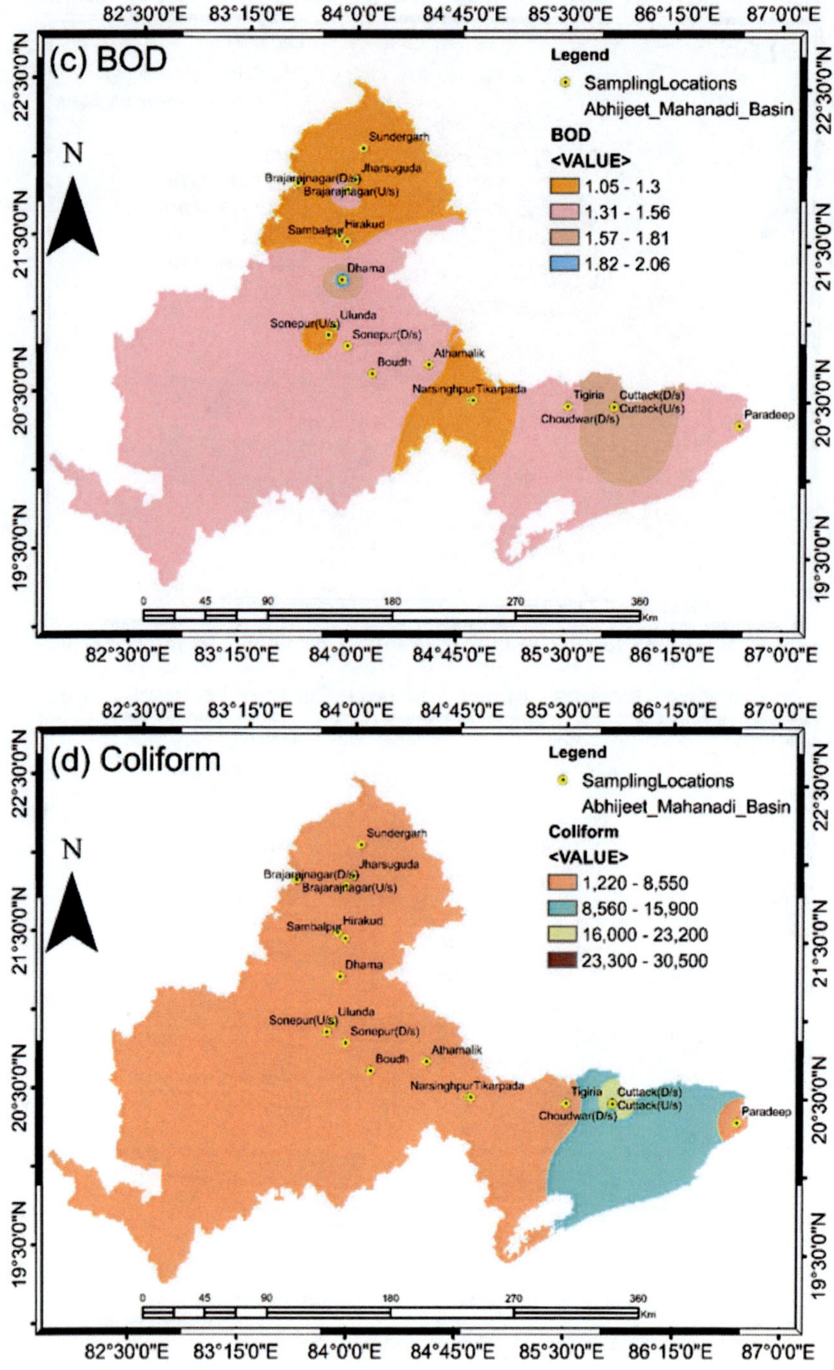

Fig. 3 (continued)

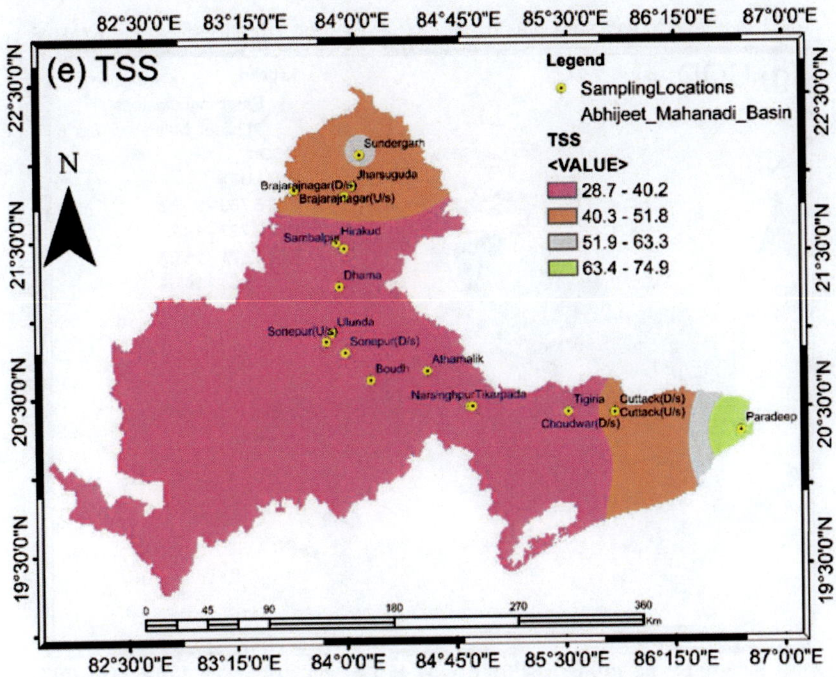

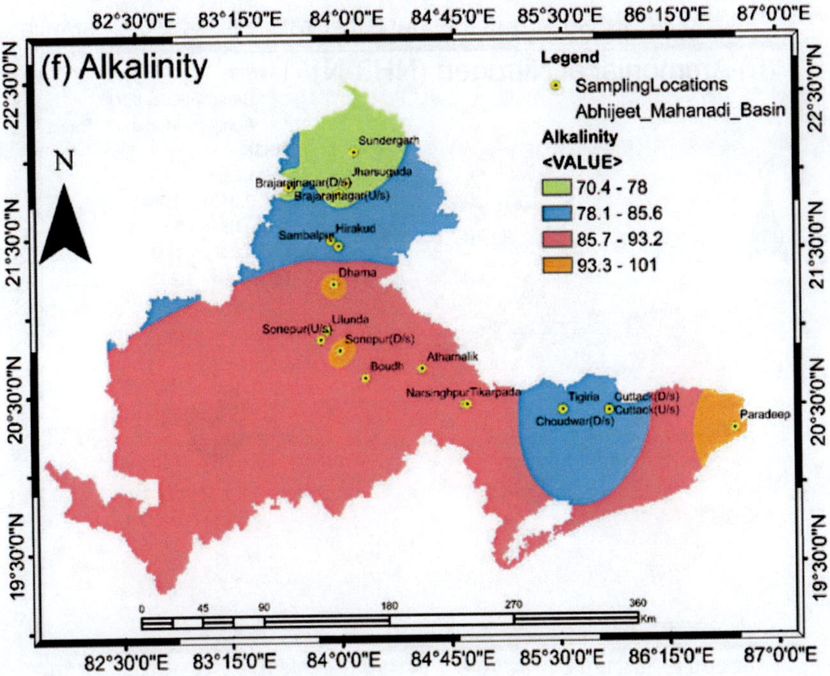

Fig. 3 (continued)

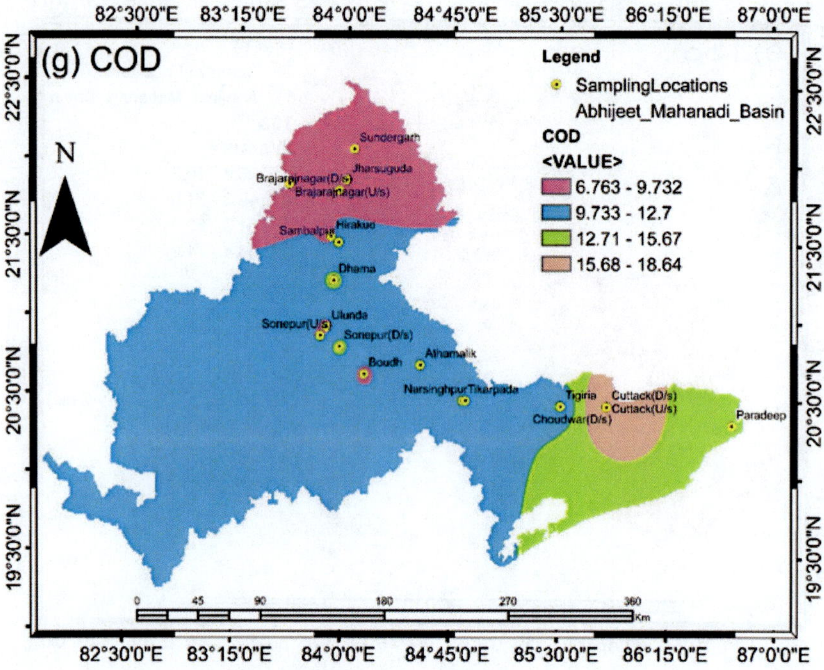

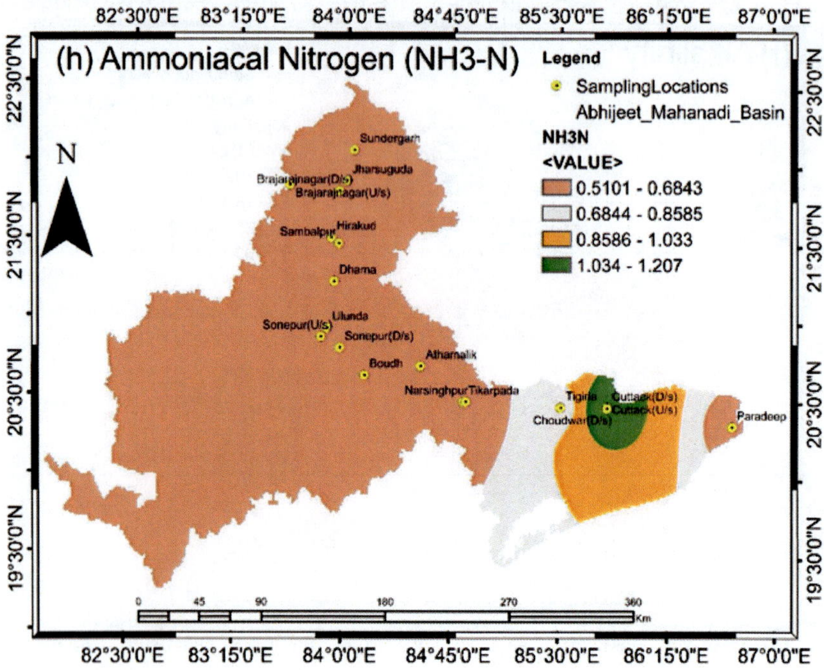

Fig. 3 (continued)

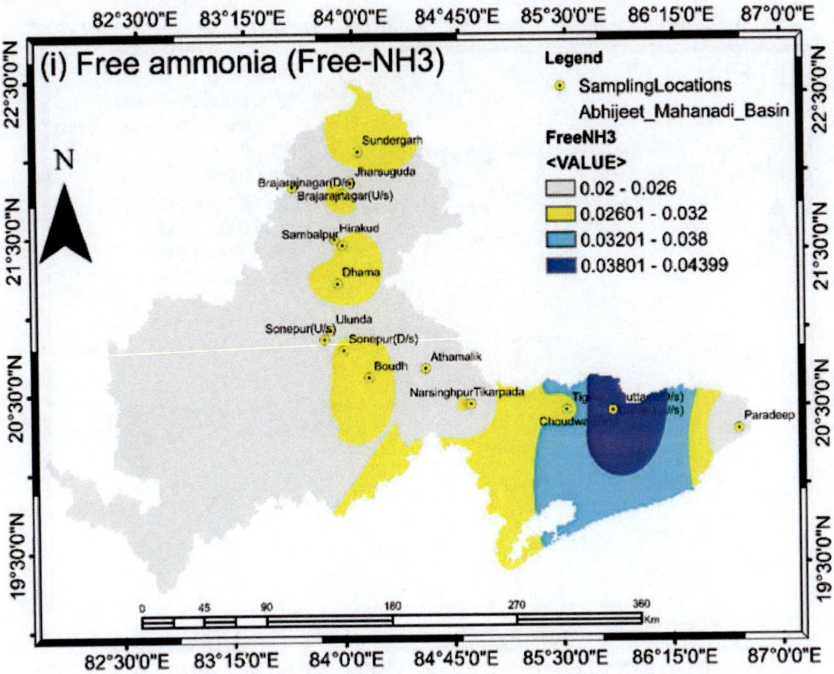

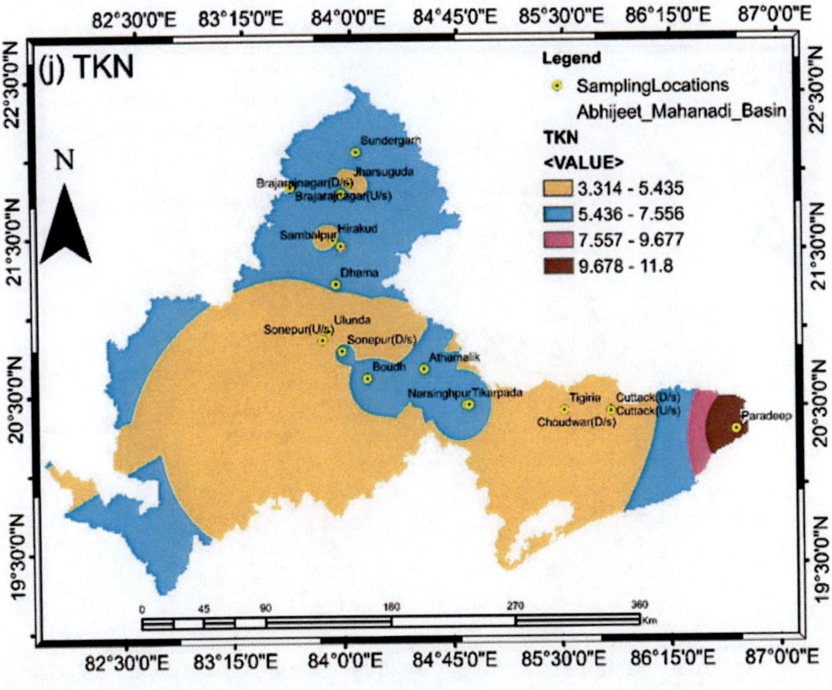

Fig. 3 (continued)

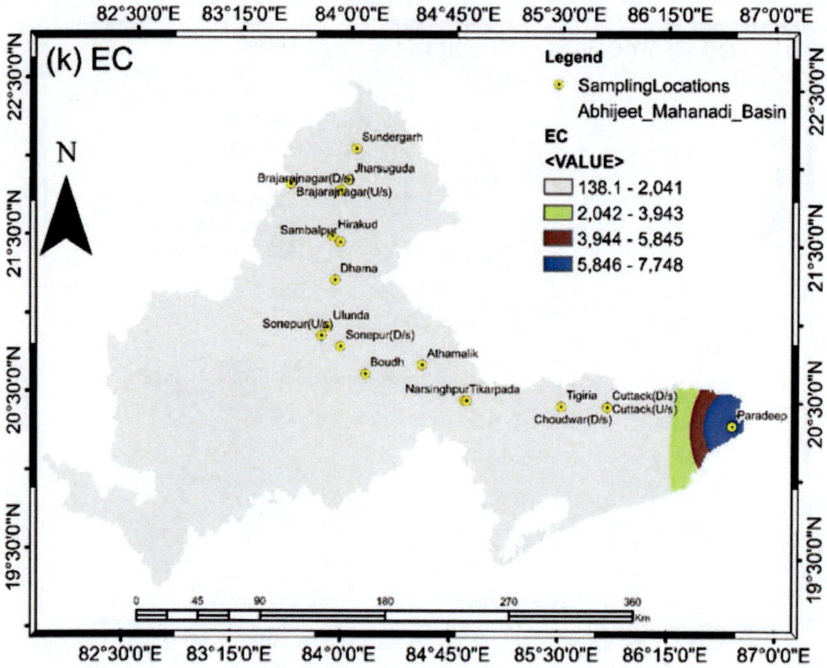

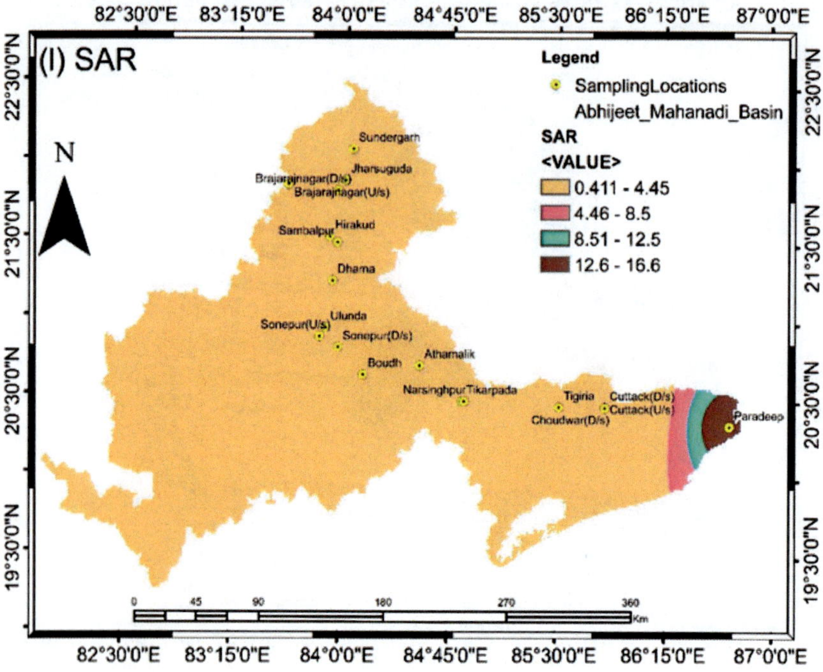

Fig. 3 (continued)

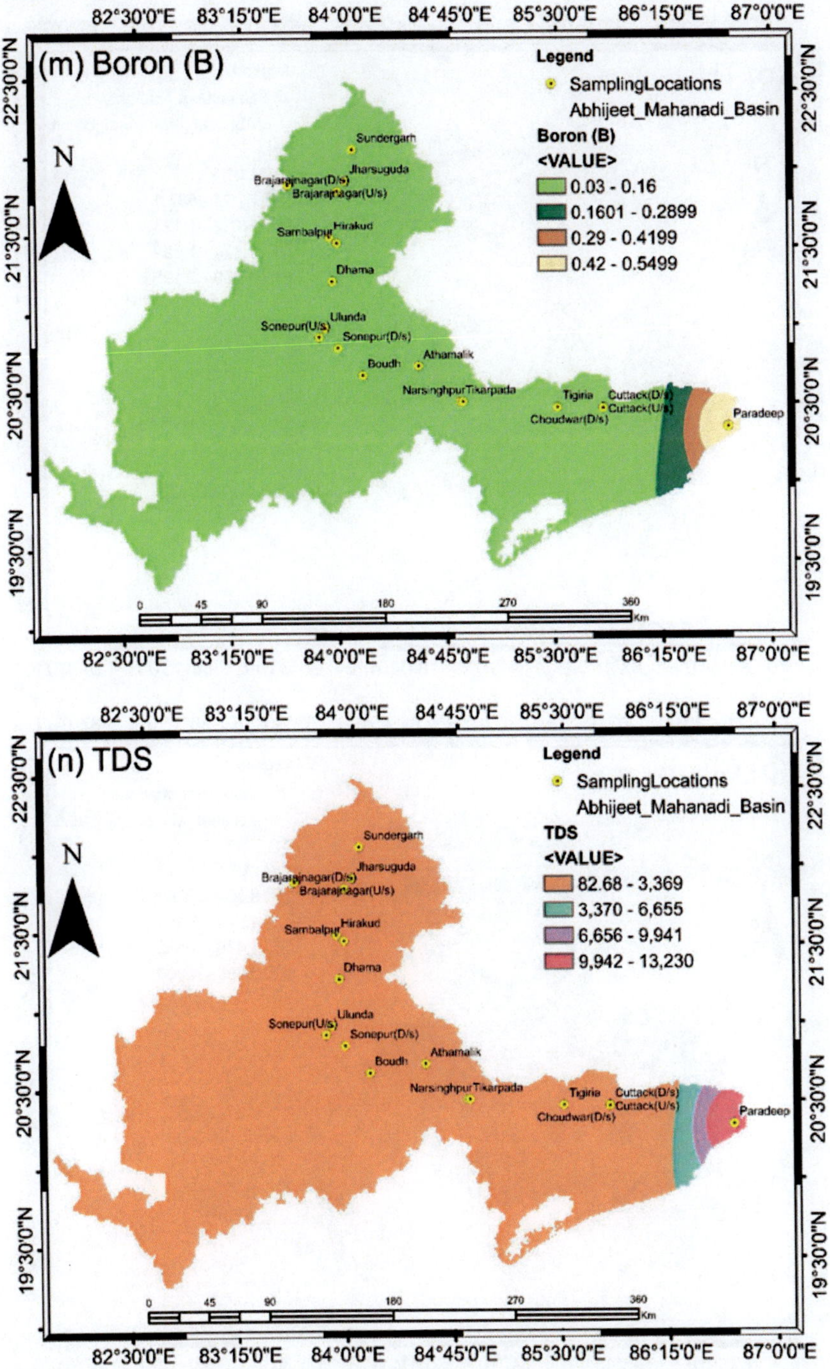

Fig. 3 (continued)

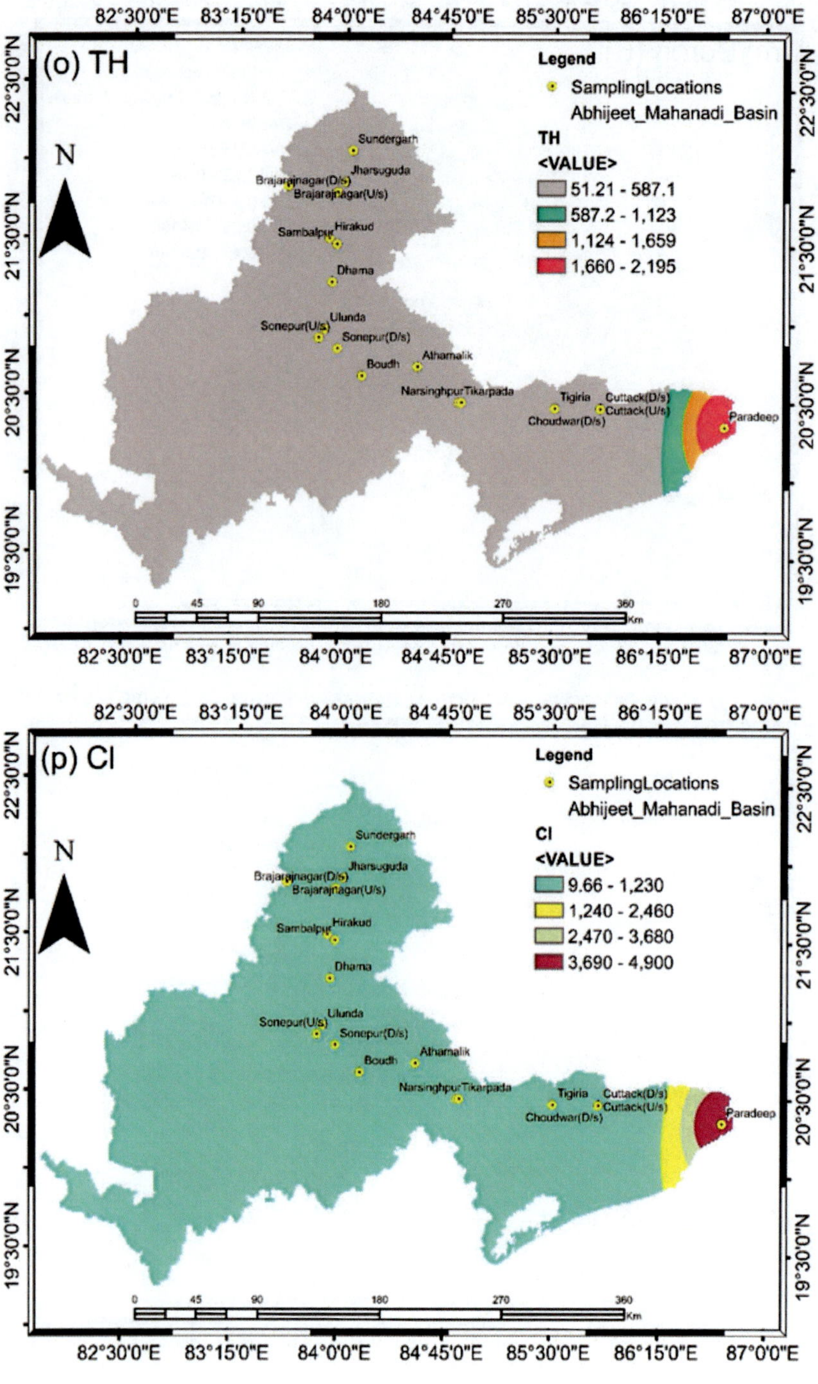

Fig. 3 (continued)

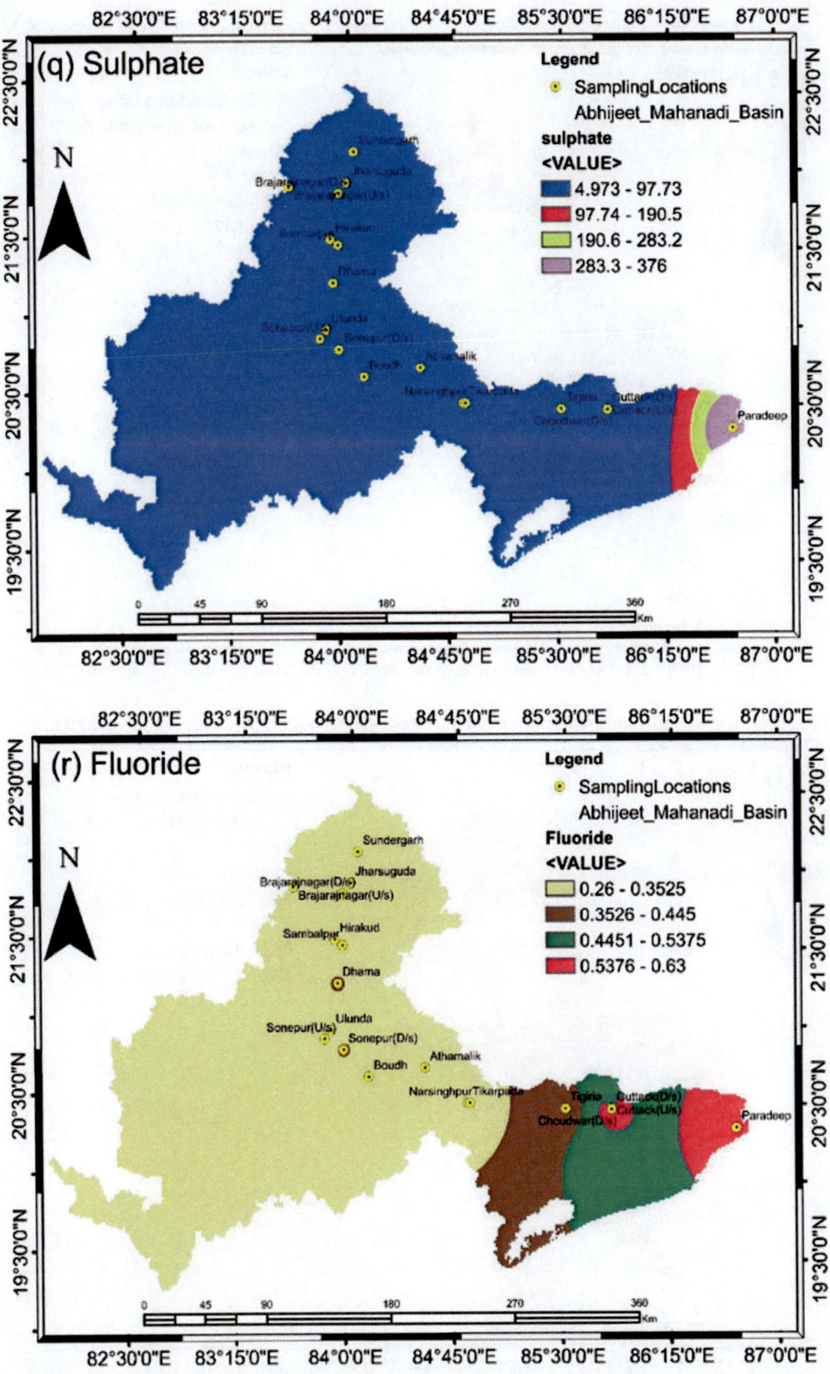

Fig. 3 (continued)

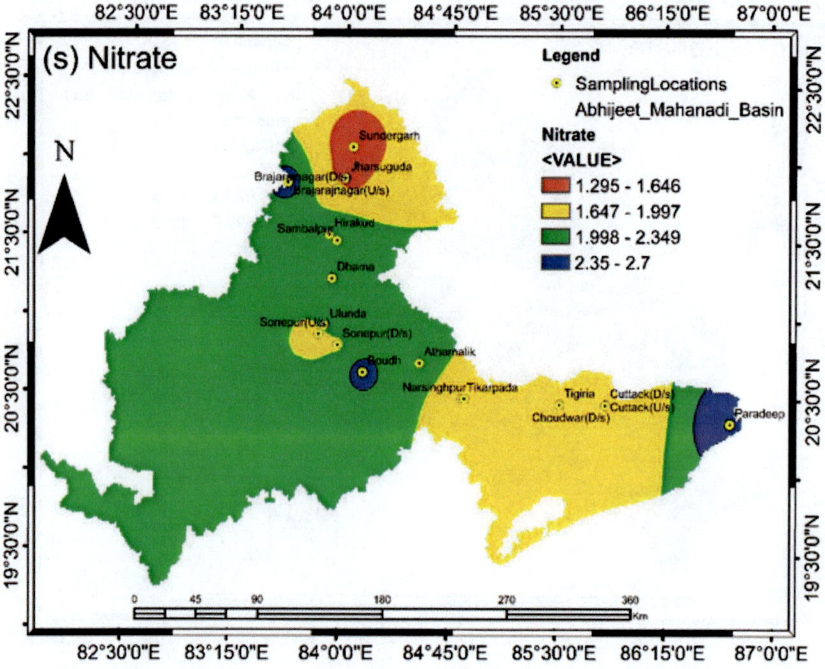

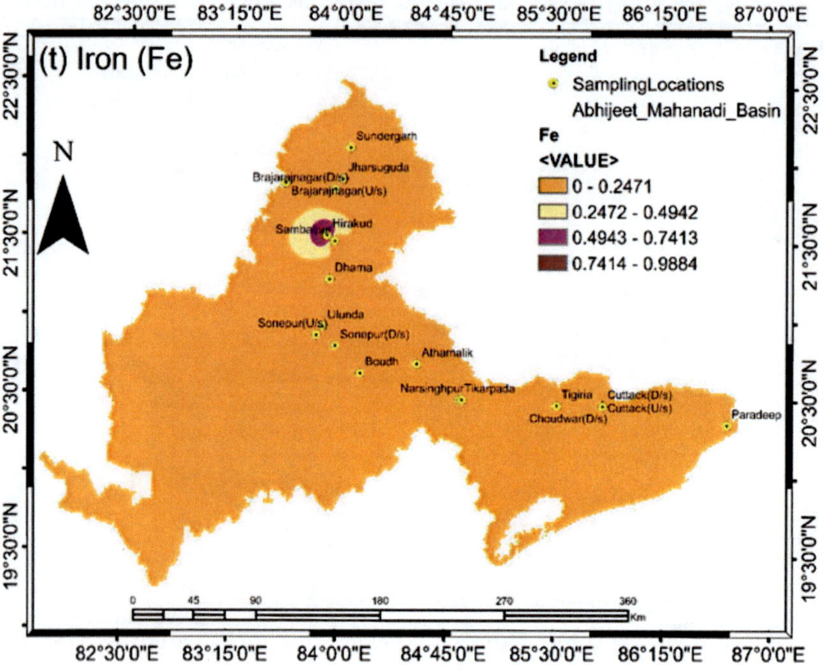

Fig. 3 (continued)

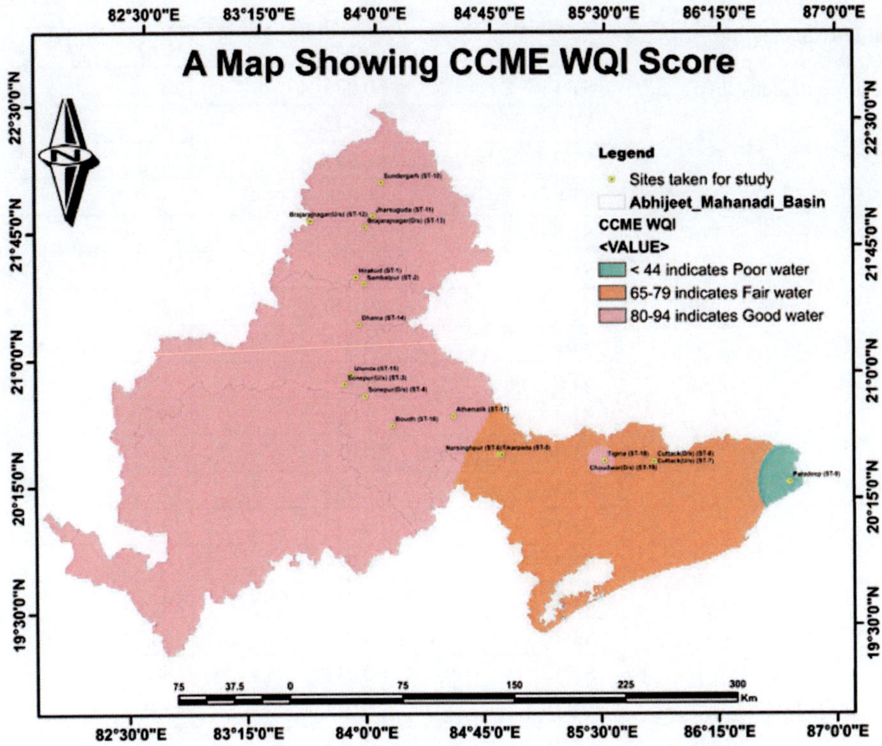

Fig. 4 CCME WQI map for the chosen study area

supplies. The agricultural portion of the research region included TKN concentrations above the standard. It might also be linked to the dissolution of solids, which causes turbidity and pollution in the water as a result of soluble chloride salts. Only one index approach is used in the current study to reach the WQI. There are other indexing methods that can be utilised if more parameters are supplied. Additionally, the current task is completed for a year. To provide a comparable outcome, future work might be implemented for seasonal periods. As the principal source of drinking water in the relevant study, the assessment of water quality is a crucial topic. Therefore, a strict management strategy must be created before the product's quality declines.

Table 1 Scores and WASPAS ranks of all the locations of study area

Locations	Scores	Rank
ST-1	0.036	18
ST-2	0.042	9
ST-3	0.037	16
ST-4	0.044	4
ST-5	0.041	11
ST-6	0.041	10
ST-7	0.040	12
ST-8	0.076	2
ST-9	0.909	1
ST-10	0.038	15
ST-11	0.035	19
ST-12	0.037	17
ST-13	0.040	13
ST-14	0.044	6
ST-15	0.038	14
ST-16	0.044	5
ST-17	0.042	8
ST-18	0.043	7
ST-19	0.063	3

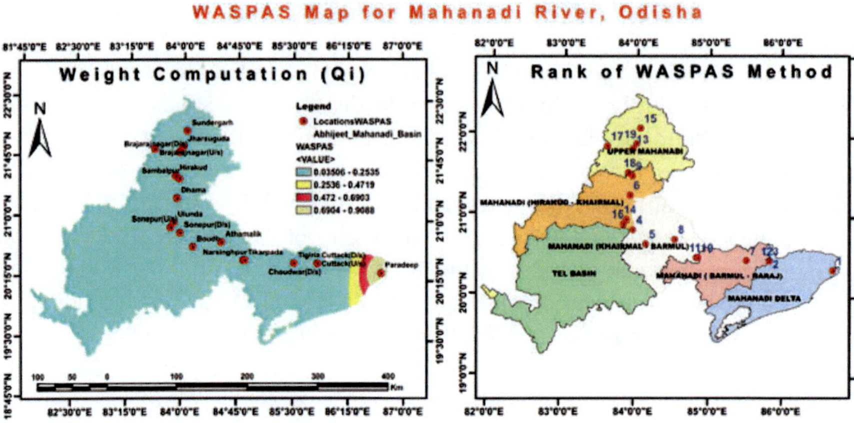

Fig. 5 Variation of WASPAS at all sampling locations

4 Conclusion

The Mahanadi River in the state of Odisha, is one of the primary water sources for irrigation and consumption. The current water shortage in the area has made it necessary to evaluate the river's water quality throughout the sample period, which is 2022–2023. To evaluate 20 physical and chemical parameters in the chosen stations, water samples were accumulated from 19 places along the river. All the spatial maps were drawn and superimposed using Arc GIS 10.8 and integrated output was obtained to represent the overall suitability of surface water for drinking purposes. The CCME WQI was initially used to evaluate the study. When combined with the multi-criteria method, the flaws caused by using just this index method were successfully fixed, giving a more complete image of the real quality of the observed water body. Overall observation advocates that surface water is slightly alkaline in nature. The parametric value for TKN exceeded the WHO permissible limit at all sites. The mean ionic dominance pattern is in the order $Fe^{2+} > B^{+}$ and $Cl^{-} > SO_4{}^{2-}$ and $NO_3{}^{-}$ for cations and anions respectively. According to the CCME WQI assessment, 73.68% of the samples have excellent quality water, making them fit for human consumption. Due to factors like seasonal volatility, unregulated fertiliser application, residential and agricultural runoffs, and agrochemicals, 10.53% of the samples have unsatisfactory grades. Based on WASPAS findings, the highest calculated coefficient was found to be 0.909, i.e. ST-9. It indicated that, in comparison to other areas, this place was the unhealthiest site. To prevent or lessen the acuteness of the cumulative pollution on less used rivers, it is necessary to build effective water wastewater treatment facilities and sewage systems because this area is prone to overuse. In some locations (ST-8, 9 and 19) along the river, this study advises the creation of a permanent commission to monitor the quality of the water resources. This study reports an exhaustive taxonomy of 20 prominent quality parameters in different categories, which will support their adoption for these models. Finally, the limitations of the current manual water quality measurement approaches are summarized to propose a set of requirements for the tech-intensive water quality profiling research and system development. Further, 'Water quality modelling' needs to be taken into consideration in order to determine how to control and limit the negative effects of industrial and human activity in the sub-catchments.

Acknowledgements The author is thankful to the Staff, C.V. Raman Global University, Bhubaneswar, Odisha, for their assistance in conducting a few of the studies. I also acknowledge the cooperation provided by the State Pollution Control Board of Odisha, India, in conducting the study. I am also appreciative of the editors' helpful criticism, which helped me to improve the manuscript. The research work is original in nature and does not contain the other findings accept referred work. Therefore, the entire manuscript is written by Abhijeet Das and it follows the ethical standards.

References

1. Yang, S., Liang, M., Qin, Z., Qian, Y., Li, M., Cao, Y.: A novel assessment considering spatial and temporal variations of water quality to identify pollution sources in urban rivers. Sci. Rep. **11**, 8714 (2021)
2. Das, A.: Multivariate statistical approach for the assessment of water quality of Mahanadi basin, Odisha. Mater. Today: Proc. **65**, A1 (2022)
3. Le, T.-H., Truong, T., Hoang, T.L., Pham, T.T.V., Pham, T.P.T., Tran, L.T.: Influences of anthropogenic activities on water quality in the Saigon river, Ho Chi Minh city. J. Water Health **20**, 491 (2022)
4. Das, A.: Water criteria evaluation for drinking purposes in Mahanadi River Basin, Odisha. In: International Conference on Trends and Recent Advances in Civil Engineering, pp. 237–260. Springer Nature Singapore, Singapore (2022)
5. Long-Ling, O., Yun-Rong, S., Jie-Qing, Y., Shu-Jie, M., Qi, Y., Yun-Long, W.: Water quality assessment and pollution source analysis of Yaojiang River Basin: a case study of inland rivers in Yuyao City, China. Water Supply **13**, 1 (2021)
6. Roy, B., Roy, S., Mitra, S., Manna, A.K.: Evaluation of ground-water quality in West Tripura, Northeast India, through combined application of water quality index and multivariate statistical techniques. Arab. J. Geosci. **14**, 2037 (2021)
7. Mahadevan, H., Krishnan, K.A., Pillai, R.R., Sudhakaran, S.: Assessment of urban river water quality and developing strategies for phosphate removal from water and wastewaters: integrated monitoring and mitigation studies. SN Appl. Sci. **2**, 772 (2020)
8. Ram, A., Tiwari, S.K., Pandey, H.K., et al.: Groundwater quality assessment using water quality index (WQI) under GIS framework. Appl. Water Sci. **11**, 1 (2021)
9. Parvin, F., Tareq, S.M.: Impact of landfill leachate contamination on surface and groundwater of Bangladesh: a systematic review and possible public health risks assessment. Appl. Water Sci. **11**, 1 (2021)
10. Teshome, M.: The Ethiopian Hearald. Berhanena Selam Printing Enterprise, LXXV (2019). Available online at: https://press.et/english/wp-content/uploads/2019/03/mar27.pdf (Accessed 10 Feb 2020)
11. Ma, T., Zhao, N., Ni, Y., Yi, J., Wilson, J.P., He, L., Du, Y., Pei, T., Zhou, C., Song, C., Cheng, W.: China's improving inland surface water quality since 2003. Sci. Adv. **6**, 3798 (2020)
12. Balamurugan, P., Kumar, P.S., Shankar, K., et al.: non-carcinogenic risk assessment of groundwater in southern part of Salem District in Tamil Nadu, India. J. Chil. Chem. Soc. **65**, 4697 (2020)
13. Zolekar, R.B., Todmal, R.S., Bhagat, V.S., et al.: Hydro-chemical characterization and geospatial analysis of groundwater for drinking and agricultural usage in Nashik district in Maharashtra, India. Environ. Dev. Sustain. **23**, 4433 (2021)
14. Brown, R.M., McClelland, N.I., Deininger, R.A., O'Connor, M.F.: A water quality index—crashing the psychological barrier. Indic. Environ. Qual., 173 (1972)
15. Sharma, A., Ganguly, R., Kumar Gupta, A.: Impact assessment of leachate pollution potential on groundwater: an indexing method. J. Environ. Eng. **146**, 05019007 (2020)
16. Saffran, K., Cash, K., Hallard, K.: CCME water quality Index 1.0 user's manual. Can. Water Qual. Guid. Prot. Aquat. Life, 1 (2001)
17. Wagh, V.M., Panaskar, D.B., Muley, A.A., Mukate, S.V.: Ground-water suitability evaluation by CCME WQI model for Kadava River Basin, Nashik, Maharashtra, India. Model. Earth Syst. Environ. **3**, 557 (2017)
18. Nath, A.V., Selvam, S., Reghunath, R., Jesuraja, K.: Ground water quality assessment based on groundwater pollution index using Geographic Information System at Thettiyar watershed, Thiruvananthapuram district, Kerala, India. Arab. J. **14**, 557 (2021)
19. Dash, S., Kalamdhad, A.S.: Discussion on the existing methodology of entropy-weights in water quality indexing and proposal for a modification of the expected conflicts. Environ. Sci. Pollut. Res. **28**, 53983 (2021)

20. Zavadskas, E.K., Baušys, R., Stanujkic, D., Magdalinovic-Kalinovic, N.: Selection of lead-zinc flotation circuit design by applying WASPAS method with single-valued neutrosophic set, Acta Montanist. Slovaca **21**, 85 (2016)
21. Zhang, Z.M., Zhang, F., Du, J.L., Chen, D.C.: Surface water quality assessment and contamination source identification using multi-variate statistical techniques: a case study of the Nanxi River in the Taihu Watershed. China Water **14**, 778 (2022)
22. Das, A.: Assessment of potability of surface water and its health implication in Mahanadi Basin, Odisha. Materials Today: Proceedings (2023)
23. Zavadskas, E.K., Turskis, Z., Antucheviciene, J., Zakarevicius, A.: Optimization of weighted aggregated sum product assessment. Elektrotechnika **122**, 3 (2012)
24. Zavadskas, E.K., Baušys, R., Lazauskas, M.: Sustainable assessment of alternative sites for the construction of a waste incineration plant by applying WASPAS method with single-Valued neutrosophic set. Sustainability **7**, 15923 (2015)
25. Ambade, B., Sethi, S.S., Kurwadkar, S., Kumar, A., Sankar, T.K.: Toxicity and health risk assessment of polycyclic aromatic hydrocarbons in surface water, sediments and groundwater vulnerability in Damodar River basin. Groundw. Sustain. Dev. **13**, 100553 (2021)
26. Das, A., Goya, A., Soni, A.: Use of water quality indices and its evaluation to verify the impact of Mahanadi river basin, Odisha. In: AIP Conference Proceedings, vol. 2721, no. 1. AIP Publishing (2023)
27. Kong, J., Ma, T., Cao, X., Li, W., Zhu, F., He, H., Sun, C., Yang, S., Li, S., Xian, Q.: Occurrence, partition behavior, source and ecological risk assessment of nitro-PAHs in the sediment and water of Taige Canal, China. J. Environ. Sci. **124**, 782 (2023)
28. Liu, J., Gao, Z., Wang, Z., Xu, X., Su, Q., Wang, S., Qu, W., Xing, T.: Hydrogeochemical processes and suitability assessment of groundwater in the Jiaodong Peninsula, China. Environ. Monit. Assess. **192**, 1 (2020)
29. Khuman, S.N., Chakraborty, P., Cincinelli, A., Snow, D., Kumar, B: Polycyclic aromatic hydrocarbons in surface waters and riverine sediments of the Hooghly and Brahmaputra Rivers in eastern and northeastern India. Sci. Total Environ. **636**, 751 (2018)
30. Wang, W., Xu, J., Qu, X., Lin, D., Yang, K.: Current and future trends of low and high molecular weight polycyclic aromatic hydrocarbons in surface water and sediments of China: insights from their long-term relationships between concentrations and emissions. Environ. Sci. Technol. **6**, 3397 (2022)
31. Zafarani, G.G., Karbalaei, S., Golshani, R., Pustokhina, I., Walker, T.R.: Baseline occurrence, distribution and sources of PAHs, TPH, and OCPs in surface sediments in Gorgan Bay, Iran. Mar. Pollut. Bull. **175**, 113346 (2022)
32. Bellanova, P., Feist, L., Costa, P.J.M., Orywol, S., Reicherter, K., Lehmkuhl, F., Schwarzbauer, J.: Contemporary pollution of surface sediments from the Algarve shelf, Portugal. Mar. Pollut. Bull. **176**, 113410 (2022)
33. Gad, M., El-Hendawy, S., Al-Suhaibani, N., Tahir, M.U., Mubushar, M., Elsayed, S.: Combining hydrogeochemical characterization and a hyperspectral reflectance tool for assessing quality and suitability of two groundwater resources for irrigation in Egypt. Water **12**, 2169 (2020)
34. Adimalla, N., Qian, H., Nandan, M.J.: Groundwater chemistry integrating the pollution index of groundwater and evaluation of potential human health risk: a case study from hard rock terrain of south India. Ecotox. Environ. Safe. **206**, 111217 (2020)
35. Elsayed, S., Gad, M., Farouk, M., Saleh, A.H., Hussein, H., Elmetwalli, A.H., Elsherbiny, O., Moghanm, F.S., Moustapha, M.E., Taher, M.A., Eid, E.M., Abou El-Safa, M.M.: Using optimized two and three-band spectral indices and multivariate models to assess some water quality indicators of Qaroun lake in Egypt. Sustainability **13**, 10408 (2021)
36. Prakash, R., Srinivasamoorthy, K., Gopinath, S., Saravanan, K., Vinnarasi, F.: Hydrogeochemical investigations to assess groundwater and saline water interaction in coastal aquifers of the southeast coast, Tamil Nadu, India. Environ. Sci. Pollut. Res. **28**, 5495 (2020)
37. Al-Gadi, K., Al-Doaan, M.: Seawater intrusion into groundwater costal aquifers at Al-Qunfudah province, western Saudi Arabia. Multi. Knowl. Electron. Compr. J. Educ. Sci. Publ., 31 (2020)

38. Das, A., Goyal, A., Soni, A.: Deciphering surface water quality for irrigation and domestic purposes: a case study in Baitarani Basin, Odisha. In: AIP Conference Proceedings, vol. 2721, no. 1. AIP Publishing (2023)
39. Laltu, A.D.: Drinking Suitability and Source Apportionment of Physical Parameters in Surface Water in Mahanadi Basin, Odisha. Odisha (April 14, 2023)
40. Morris, B.L., Lawrence, A.R., Chilton, P.J., Adam, B., Caylow, R.C., Klink, B.A.: Groundwater and Its Susceptibility to Degradation: A Global Assessment of the Problems and Options for Management; UNEP Early Warning and Assessment Report Series RS. 03–3; United Nations Environment Programme: Nairobi, Kenya, 2003 (2020)

Water for Food Drying: Geothermal Energy-Based Food Dryer

Namrata Bist, Roshni Kumari, Kriti Yadav, and Anirbid Sircar

Abstract Different drying methods are practised worldwide for the preservation and longevity of herbs and other ingredients. The present work describes the concept of utilising renewable energy in the form of water, i.e. geothermal energy for food drying. In the current study, the notion is introduced and articulated for the first time in India. This study describes the demonstration of geothermal energy-based food dryer established in the Dholera region of Gujarat and its application in the drying of Ocimum sanctum (tulsi) leaf. The instrument is efficiently designed such that there are no hazardous emissions or extensive energy requirements for drying materials. Moreover, its uniqueness lies in the fact that low-temperature geothermal energy can be utilised as an energy source, and the drying conditions can be monitored as desired. Also, there is a uniform distribution of heat during drying, which helps to retain the structural behaviour of the products after drying. The investigation of moisture content and phytochemical analysis of geothermally dried tulsi leaf and its comparative study with other drying methods were done to evaluate the system's integrity. The moisture content and phytochemicals of the geothermally dried leaf were found 16.6% and showed the presence of eugenol, camphene hydrate, farnesene, amorphene, etc. which are responsible for the medicinal and therapeutic significance of the leaf. Thus, enhanced preservation time and retention of essential phytoconstituents are justified through this technology.

Keywords Geothermal water · Food dryer · Water drying · Tulsi leaf · Phytochemicals

---

N. Bist (✉) · R. Kumari · K. Yadav · A. Sircar
Pandit Deendayal Energy University, Raisan, Gandhinagar, Gujarat 382007, India
e-mail: namrata.bist@spt.pdpu.ac.in

S. Saxena et al. (eds.), *Emerging Materials and Technologies in Water Remediation and Sensing*, Lecture Notes in Civil Engineering 439,
https://doi.org/10.1007/978-981-99-6762-9_3

1 Introduction

There are several drying methods used in the food and herbal industry for drying products like oven drying, microwave drying, sun drying, etc. [1, 2]. Drying of substances is practised to increase the longevity and preservation time of food products since the moisture content is mainly responsible for microbial growth hence drying or reduction in moisture content prevents or delays the spoilage of products [3–5]. All these conventional methods are limited by one or more shortcomings like non-renewable fuel sources or time consumption, or economic liabilities. The geothermal energy-based drying method has several advantages over other conventional methods the source of energy utilised is renewable and sustainable, and the input power can be monitored. A new alternative and sustainable approach to the geothermal-based food drying method is introduced and discussed in the present work for the first time in the country. Dholera holds the potential viability for geothermal exploration and exploitation in terms of hot water springs and site suitability for plant setup as well as industrial accessibility [6–8]. With the space heating and cooling system set up at Dholera, geothermal water can attain temperatures of about 75–80 °C. This outlet water temperature is used for honey processing and heating purposes like drying [7, 8]. These inspired the new technology development for applying geothermal energy in food and herbs drying, which could emerge as a sustainable model in drying technologies in the future. Ocimum sanctum, or Holy Tulsi, is a well-known medicinal plant, and its leaf is one of the most significant parts responsible for its therapeutic and medicinal importance. A large number of organic phytoconstituents and secondary metabolites are present in the leaf, which forms a rich source of medicinal drugs and ayurvedic formulations. These drugs are helpful for the treatment of many diseases, including malaria, bronchitis, hepatitis, asthma, allergy, arthritis, dysentery, chronic diseases, skin diseases, eye diseases, gastric problems and several others [9–11]. Moreover, the essential oils and phytochemicals in them show antifungal, anti-allergic, anti-inflammatory, antibacterial, antifertility, anti-microbial, anti-diabetic and anti-cancerous behaviour [12–17]. The present work describes the new concept, working and demonstration of geothermal energy based food dryer established in the Dholera region of Gujarat and its application in the drying of tulsi leaf, an important medicinal constituent. In addition, the study of moisture content and phytoconstituents of geothermally dried tulsi leaf is also explained here to justify the application of the method on a real ground basis. The work also conglomerates a comparative discussion on the effect of geothermal drying on other drying methods on the moisture content of leaf samples.

2 Instrumentation

2.1 Geothermal Energy-Based Food Dryer

The geothermal-based food dryer is established in the Dholera region of Gujarat in India. The majority of food dryers in the world uses high energy heat sources for catering to the low heat requirements of food drying. This setup is efficient as it uses renewable energy which is hot geothermal water as a source of energy to generate heat for the drying of products. The setup was demonstrated for drying a medicinal herb, *Ocimum sanctum* (tulsi leaves) using geothermal hot water as the heating source. The schematic of the instrument with figure (Fig. 1.) is mentioned below.

The Geothermal energy-based food dryer consists of two main units, viz., (i) mechanical unit, (ii) drying chamber, and a flexible insulated duct. The mechanical unit consists of an electric air heater, heat exchanger, air blower, flow control valves, rotameter, control panel and four temperature sensors. The control panel comprises the main switch, blower on/off switch, heater on/off switch, flow regulator, temperature scanner and temperature controller.

The drying chamber consists of six drying trays. This unit has 40 mm insulation inside the three walls to maintain the internal temperature. The door is insulated with appropriate gaskets to prevent any kind of leakage. A hot air duct connects the mechanical unit and the drying chamber. Table 1 represents the different types of insulating materials and their thermal properties.

The system is initialised with high-temperature geothermal water entering the mechanical unit through the inlet flow control valve, which then passes through the rotameter. Simultaneously, atmospheric air is sucked by the air blower through the air filter. The geothermal water and air enter into two fin and tube type heat

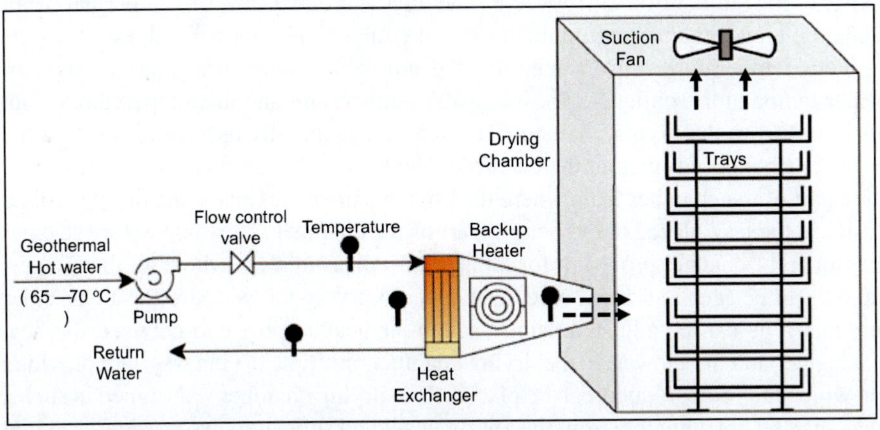

Fig. 1 Schematic diagram of the geothermal energy-based food dryer

Table 1 Types of insulating materials and their thermal properties

Insulation material	Thermal conductivity W/MK-less indicates better performance	Thermal resistance M^2K/W-more indicate better performance	U-value W/M^2K
Glass fibre	0.032–0.044	3.10–2.25	0.32–0.44
Rock fibre	0.035–0.044	2.85–2.25	0.35–0.44
Sheep's wool	0.042	2.38	0.42
Expanded polystyrene (EPS)	0.036	2.77	0.36
Multi-foils	0.040	–	–
Hemp	0.039	2.56	0.39
Extruded polystyrene (XPS)	0.028–0.036	3.44–2.77	0.29–0.0.36
Polyurethane foam board (PUR)	0.22–0.29	0.45–3.44	2.22–0.29
Polyisocyanurate foam board (PUR)	0.021–0.022	4.76–4.54	0.21–0.22
Phenolic foam board	0.021	4.76	0.21
Evacuated panels	0.004	5.00	0.20
Aerogel board	0.013	0.77	1.29

exchangers connected in series. The temperature of the geothermal water entering the heat exchanger ranges from 45 to 70 °C depending on the potential of the geothermal spring. The air encompasses the chamber with lower temperatures ranging from 25 to 40 °C depending on atmospheric conditions therein. As a result of heat exchanges between these two, the temperature of the air rises, which is drawn by the blower and is supplied to the drying chamber through the insulated duct. The outlet water exists in the outlet flow control valve. The inlet and outlet flow of geothermal water could be monitored per requirement using the inlet and outlet ball valves.

Four temperature sensors measure the ambient temperature, inlet geothermal water temperature, outlet geothermal water temperature and the temperature of hot air supplied to the drying chamber. The sensor is generally used to understand the heat distribution throughout the chamber. The heat will be higher at the bottom and side part of the chamber from where the hot air will enter. Hence, the drying will be faster in the tray placed at the bottom part of the chamber. The temperature scanner has an eight-channel provision for temperature measurement where all the temperatures can be seen and monitored. The specific drying temperature for a particular product type can be adjusted using backup air heaters provided between the heat exchanger and the blower. In the drying chamber, multiple drying trays are provided on which the food products can be placed. This drying chamber is designed in such a way that the hot air enters from the bottom level and splits into two streams such that the chamber is uniformly heated. The hot air absorbs the moisture content from the products and ejects out from the exhaust provided on the top surface of the drying chamber.

2.2 Gas Chromatography-Flame Ionisation Detector (GC-FID)

Tulsi leaves are of great medicinal significance, and hence their preservation is a necessity. So, the drying of tulsi leaves is investigated under optimum conditions (collection of fresh natural plant leaves without any chemical modification or pretreatment at standard temperature and pressure conditions). Fresh leaves were collected from the local tulsi plant. The pretreatment was done to remove dust and other particles. The leaves were separated from the stalk manually and after proper separation and cleaning, these were packed and kept for the drying experiments. The moisture content forms the major portion of a leaf as it is involved in holding all the components as a solvent moiety or being present as hydrated groups. The microbial activity requires moisture for its growth and is responsible for its spoilage. Therefore the preservation in terms of reduction in moisture content and retention of phytoconstituents in tulsi leaves are of utmost importance. Hence, this is chosen for the application-based study of geothermal energy food dryers. The phytochemical analysis of geothermally dried leaves was done by GC FID (Agilent Technologies 7890A GC System). Among several well-known solvents for the extraction of phytoconstituents, methanol has been identified as the most effective solvent for the extraction of phytochemicals in terms of high yield, the present study used methanol as a solvent for extraction purposes [18]. To understand the properties of antioxidant compounds from plants generally methanol and ethanol extraction is applied. The methanolic extract of the dried sample was analysed for the phytoconstituent study. The composition of the dried sample concerning the retention time and peak percent was obtained. The details of the gas column chromatography are given in Table 2. The oven (METALAB) is used for all the drying experiments.

Table 2 GC-column details

Parameters	Value
Flame	Hydrogen
Carrier gas	Nitrogen
Column length	30 m
Column diameter	0.32 mm
Film	0.25 μm
Temperature	20–265 °C
Solvent	Methanol

3 Material and Methods

3.1 Drying Experiments

The moisture content of the fresh leaves was found using the proximate analysis method, which involves heating the sample for a certain period. A measured amount of the leaves were taken and placed in the oven at 110 °C for 5 min. The loss in weight after heating was obtained, and moisture percent was calculated using the formula below

$$\%Moisture = [(W_2 - W_1) - (W_3 - W_1)] \times \frac{100}{(W_3 - W_1)} \tag{1}$$

where W_1 = Weight of empty metallic dish, W_2 = Weight of metallic dish with the sample, W_3 = Weight of metallic dish with the dried sample.

The moisture content after applying different drying methods was calculated using the same formula. All the experiments are performed in duplicate sets for better accuracy of results. The different drying methods involved are discussed in detail.

Oven Drying. A measured amount of (5 gm) of pre-treated leaves was placed in a heating crucible in an oven in two sets, one for 20 min at 75 ± 5 °C and the other for 8 h at 75 °C. The weight of the dried leaves along with the crucible in each case is noted down, and the moisture content or moisture loss is calculated using Eq. 1.

Sun Drying. A flat plate containing the known weight of leaves was placed in direct sunlight for drying at an outside temperature of 35 °C for 6 h. To monitor the drying process, the initial and final outside temperatures are noted down and found around 35 °C. The loss in weight after drying is measured and moisture content is calculated by Eq. 1.

Air Drying. 5 gm of the sample was weighed and kept in a plate for air drying in a ventilated dark room. The material was kept for air drying at an optimum room temperature (26 °C) (measured temperature, without any modification) for 5 days. The temperature of the air drying method was kept as measured with no external modification so that the most common and simple mode of traditional drying might be compared with the other methods. The loss in weight is again measured, and moisture percent was calculated using Eq. 1.

Geothermal Drying. The moisture content of the dried leaf was done using the proximate analysis method, which involves heating the sample for a certain period. The pre-treated fresh tulsi leaves were placed in the geothermal dryer for 6–8 h at 75 ± 5 °C in a measured amount (1 g). After drying, weight loss was noted down, and moisture percent was calculated.

4 Result and Discussion

4.1 Moisture Content Analysis

The moisture content of fresh tulsi leaves was found to be 92.26% ± 0.5. All the experiments are performed in duplicate sets for better accuracy of results. This content shows that the leaves form a highly perishable part of the plant and are the storehouse for most of the phytoconstituents that are bound to water molecules in one or the forms like the water of crystallisation, hydrated salts or as solvent moieties. The moisture content obtained for sun-dried leaves was 21.42% ± 1 which might be due to the low-temperature intensity of natural sun drying applied for a short time interval of day-time. The moisture content obtained for Oven dried leaves at 100 °C for 20 min and 75 °C for 8 h were 20.26% ± 1 and 18.89 ± 1 respectively. The high-temperature treatment of 100 °C resulted in a better reduction in moisture percent than the sun drying method. Moreover, for better interpretation of the comparative analysis of different drying methods, the experimental conditions for drying were kept constant for both oven and geothermal dying, and the results showed more effective moisture reduction in the case of geothermal drying. This might be attributed to the fact that geothermal drying caused uniform heat distribution that resulted in greater moisture loss in the sample. In the air-drying method, the sample was left in a free air-drying atmosphere in a room for a very long interval of time, as mentioned earlier for moisture loss. The moisture content obtained for the free air drying method was 19.13 ± 1%. The high loss in moisture content might be attributed to the fact that longer drying time exposure leads to the conduction of heat to the interior layers from the surface, as a result of which the rate of surface evaporation exceeds the rate of diffusion to the leaf surface [2, 19].

The moisture content found from geothermal drying was about 16.66% which is the minimum compared to other drying methods like oven drying, air drying, sun drying, etc. Moreover, for more accuracy, the comparative analysis of geothermal drying with oven drying at constant experimental conditions was done which also ensured better moisture removal in the case of geothermal drying that in turn justified the design setup. The low value signifies the efficiency of geothermal energy as a new, convenient and sustainable alternative for drying materials as compared to other existing methods of drying. This method hence can be a better alternative for drying a variety of food, herbs and other ingredients. The comparative effect of geothermal drying to other drying methods on the moisture percent of tulsi leaves is illustrated in Fig. 2 and Table 3.

4.2 Phytoconstituent Analysis

Tulsi leaf consists of a diverse group of phytochemicals, including polyphenols, terpenes, alkaloids, sesquiterpene hydrocarbon, aromatic compounds, esters, etc.

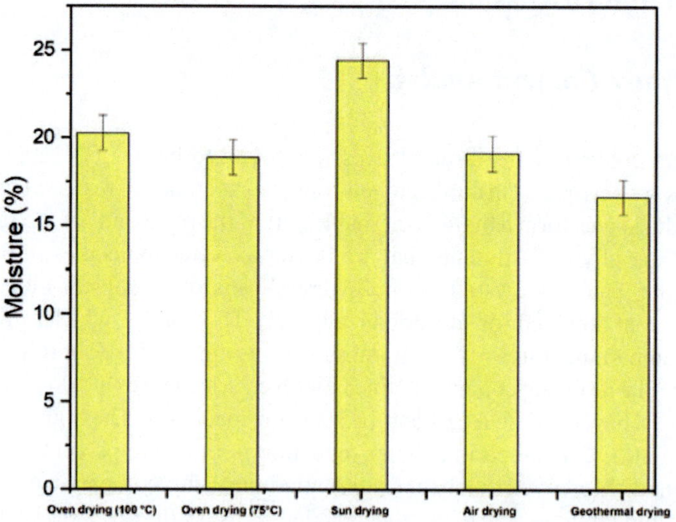

Fig. 2 Moisture percent of dried tulsi leaves using different drying methods with error bars

Table 3 Comparative effect of geothermal drying to other drying methods on the moisture content of tulsi leaves

Method of drying for sample tulsi leaf	Parameters	Moisture percent (%)
Oven drying	100 °C, 20 min	20.26 ± 1
Oven drying	75 °C, 8 h	18.89 ± 1
Sun-drying	35 °C, 6 h	24.42 ± 1
Air drying	26 °C, 10 days	19.13 ± 1
Geothermal drying	75 °C, 8 h	16.66 ± 1

that are responsible for various biological activities of this herb, including anticancer, anti-inflammatory, anti-microbial, anti-allergic, anti-diabetic, antitumor and several other therapeutic properties [20]. In addition to its direct application in medicinal drug synthesis, it has other significant benefits like wound healing, soothing, antioxidant, and flavouring agents in food, perfumes, and healthy beverages. This herb's pharmacological and toxicological efficiency make it a potential curative agent for human health and diseases. Therefore, its preservation and longevity by a convenient method are of utmost importance. Moreover, the retention of the phytochemicals responsible for their pharmacological activities by different preservation methods is an important requirement. Thus, the methanolic extract of geothermally dried leaves is tested for the presence of valuable phytochemicals with the help of gas column chromatography through the flame ionisation technique. The results obtained for some notable compounds with their retention time, peak percent and function are illustrated in Table 4.

Table 4 Phytoconstituents of geothermally dried tulsi leaves

Components	RT (min.)	% Peak	Functions	References
Camphene hydrate	15.16	0.31	Anti-microbial, anti-inflammatory, anticancer, respiratory and cardiovascular disorders	Bhattacharjya et al. [20–22]
Cyclohexane triethenyl	13.50	0.22	Anti-inflammatory, antibacterial, anti-endemic, antispasmodic, anti-microbial, antioxidant	Borah et al. [9]
Eugenol	18.99	0.40	Analgesic, anaesthetic, antibacterial, anti-inflammatory, anti-endemic, antioxidant, antiviral, antitumor, antiulcer	Borah et al. [9, 21]
Farnesene	22.20	0.51	Allergenic, analgesic, anaesthetic, antibacterial, anti-inflammatory, anti-endemic, antioxidant, antiviral, antitumor, antiulcer	Borah et al. [9, 20]
Amorphene	23.52	1.10	Pro-oxidant treats disorders related to oxidative stress	Yousif et al. [19, 21]

* RT=Retention time (min.)

The table above shows the essential phytoconstituents with all the medicinal and therapeutic significance of the leaf. Eugenol, with a peak percent of 0.40, is an oxygenated monoterpenoid and is a very essential component since it has anti-fungal, antibacterial, anti-cancerous, cardiovascular activities and several others. It also significantly plays a role in treating respiratory ailments and oxidative stress [23–27]. In addition to this, the anti-inflammatory, antioxidant and vasodilator properties of this phytoconstituent serve as an effective medium for cardiovascular protection as well as the treatment of stress-related disorders [20, 28].

Farnesene is another important phytoconstituent present in the dried leaf, with a peak percent of 0.51 and a retention time of 22.20. It is responsible for the anti-microbial, anaesthetic, anti-inflammatory and anti-allergic properties of the specimen. This, along with other phytoconstituents like eugenol, linalool, caryophyllene, etc. shows strong antifungal activities [24].

Cyclohexane triethenyl is another known phytocomponent showing properties like anti-inflammatory, antioxidant, anti-microbial, anti-endemic and antispasmodic. Antispasmodic is a property of showing relief to spasms or tummy pain. It is present in the dried leaf at a peak percent of 0.22 with a retention time of 13.5 [9, 20].

Amorphene is a sesquiterpene hydrocarbon in the geothermally dried leaf with a percent peak at 23.52 and a retention time of 1.10. It acts as a pro-oxidant and shows activities like relieving oxidative stress or helping to treat disorders related to oxidative stress along with other sesquiterpene hydrocarbons like cadinene, selinene, etc. [21]. Camphene hydrate is found in the dried leaf with a peak percent of 0.31 and a retention time of 15.16. It is a monoterpene hydrocarbon with numerous pharmacological properties like anticancer, anti-inflammatory, anti-microbial, and respiratory and cardiovascular disorders. There are numerous other phytoconstituents in variable concentrations and different retention times that are present in the dried sample, like camphor, limonene, carvacrol, humulene, bisabolol, cardenene, caryophyllene, ocimene, terpinene, citral, thymol, etc. that along with other constituents are responsible for therapeutic and medicinal importance of the leaf specimen.

5 Conclusion

The geothermal energy-based food dryer at the Dholera region of Gujarat was successfully demonstrated for drying tulsi leaf, a universal medicinal specimen. The moisture percent obtained for geothermal drying was about 16.66%. The large difference in moisture content before and after geothermal drying signifies the potential for dehydration and enhanced preservation time of the specimen through a renewable energy resource. Generally, the dried tulsi leaves have a life expectancy of more than six months which depends on the percentage moisture content of the leaves. Since the research articles aim to estimate the moisture content which was found to be minimum as compared to other methods, it can be suggested that it has enhanced preservation time. Moreover, the value of moisture percent obtained after geothermal drying is minimum when compared to other drying methods like oven drying, air drying, sun drying, etc. This low value supports the efficiency of the geothermal-based drying method to be a new, convenient and sustainable alternative for drying materials over and above other existing drying methods. The study reveals that the various phytoconstituents or secondary metabolites essential for various medicinal and therapeutic values of tulsi leaf, like eugenol, camphene hydrate, farnesene, amorphene and cyclohexenyl triethenyl are retained or present in significant peak percent in the geothermally dried leaf which also remarks the integrity of the process of drying through this method. Therefore, this method can be a better alternative for drying various food, herbs, and other ingredients and would be a sustainable tool for driving drying technology based on sensitive and limited drying methods.

References

1. Alp, D., Bulantekin, Ö.: The microbiological quality of various foods dried by applying different drying methods: a review. Eur. Food Res. Technol. **247**, 1333–1343 (2021)
2. Danso-Boateng, E.: Effect of drying methods on nutrient quality of basil (Ocimum viride) leaves cultivated in ghana. Int. Food Res. J. **1**;**20**(4) (2013)
3. Parmar, M.R., Bhalodiya, V.B., Kapdi, S.S.: Temperature effect on drying and phytochemicals of basil leaves. Int. J. Eng. Sci. Investig. **7**(1), 34–44 (2018)
4. Taheri-Garavand, A., Rafiee, S., Keyhani, A., Javadikia, P.: Modeling of basil leaves drying by GA–ANN. Int. J. Food Eng. **16**;**9**(4), 393–401 (2013)
5. Pirbalouti, A.G., Mahdad, E., Craker, L.: Effects of drying methods on qualitative and quantitative properties of essential oil of two basil landraces. Food Chem. **1**;**141**(3), 2440–9 (2013)

6. Sircar, A., Shah, M., Sahajpal, S., Vaidya, D., Dhale, S., Chaudhary, A.: Geothermal exploration in Gujarat: case study from Dholera. Geothermal Energy. **3**(1), 1–25 (2015)
7. Yadav, K., Sircar, A.: Application of low enthalpy geothermal fluid for space heating and cooling, honey processing and milk pasteurization. Case Stud. Therm. Eng. **1**; **14**, 100499 (2019)
8. Sircar, A., Yadav, K., Shah, M., Sahajpal, S., Chinan, R.: Seismic imaging for geothermal exploration: a case study from Dholera, India. Peninsula. **3**(7) (2016)
9. Borah, R., Biswas, S.P.: Tulsi (Ocimum sanctum), excellent source of phytochemicals. Int. J. Environ., Agric. Biotechnol. **3**(5), 265258 (2018)
10. Joshi, S.V., Bothara, S.B., Surana, S.J.: Evaluation of aqueous extract of Ocimum sanctum in experimentally induced Parkinsonism. J. Chem. Pharm. Res. **3**(1), 478–487 (2011)
11. Wei, A., Shibamoto, T.: Antioxidant/lipoxygenase inhibitory activities and chemical compositions of selected essential oils. J. Agric. Food Chem. **23**;**58**(12), 7218–25 (2010)
12. Filip, S., Vidović, S., Vladić, J., Pavlić, B., Adamović, D., Zeković, Z.: Chemical composition and antioxidant properties of Ocimum basilicum L. extracts obtained by supercritical carbon dioxide extraction: drug exhausting method. J. Supercrit. Fluids. **1**;**109**, 20–5 (2016)
13. Farouk, A., Fikry, R., Mohsen, M.: Chemical composition and antioxidant activity of Ocimum-basilicumL. Essential oil cultivated in Madinah Monawara, Saudi Arabia and its comparison to the Egyptian chemotype. J. Essent. Oil Bear. Plants. **19**, 1119–1128 (2016)
14. Falowo, A.B., Mukumbo, F.E., Idamokoro, E.M., Afolayan, A.J., Muchenje, V.: Phytochemical constituents and antioxidant activity of sweet basil (Ocimum basilicum L.) essential oil on ground beef from boran and nguni cattle. Int. J. Food Sci. (2019)
15. Tewari, D., Pandey, H.K., Sah, A.N., Meena, H., Chander, V., Singh, R., Singh, P.: Phytochemical, antioxidant and antidepressant evaluation of Ocimum basilicum, O. tenuiflorum, O. kilimandscharicum grown in India. J. Biol. Act. Prod. Nat. **4**;**5**(2), 120–31 (2015)
16. Londhe, A.M., Kulkarni, A.S., Lawand, R.V.: In-vitro comparative study of antibacterial and antifungal activities: A case study of Ocimumkilimandscharicum, Ocimumtenuiflorum and Ocimumgratissimum. Int. J. Pharmacogn. Phytochem. Res. **7**, 104–110 (2015)
17. Pandey, S.: Antibacterial and antifungal activities of OcimumgratissimumL. Int J Pharm Pharm Sci **9**, 26–31 (2017)
18. Truong, D.H., Nguyen, D.H., Ta, N.T., Bui, A.V., Do, T.H., Nguyen, H.C.: Evaluation of the use of different solvents for phytochemical constituents, antioxidants, and in vitro anti-inflammatory activities of Severinia buxifolia. J. Food Qual. (2019)
19. Yousif, A.N., Scaman, C.H., Durance, T.D., Girard, B.: Flavor volatiles and physical properties of vacuum-microwave-and air-dried sweet basil (OcimumbasilicumL.). J. Agric. Food Chem. **47**, 4777–4781 (1999)
20. Bhattacharjya, D., Adhikari, S., Biswas, A., Bhuimali, A., Ghosh, P., Saha, S.: Ocimum phytochemicals and their potential impact on human health. In: Phytochemicals in Human Health (2019)
21. Yamani, H.A., Pang, E.C., Mantri, N., Deighton, M.A.: Antimicrobial activity of tulsi (Ocimum tenuiflorum) essential oil and their major constituents against three species of bacteria. Front. Microbiol. **17**; **7**, 681 (2016)
22. De Lima, V.T., Vieira, M.C., Kassuya, C.A., Cardoso, C.A., Alves, J.M., Foglio, M.A., De Carvalho, J.E., Formagio, A.S.: Chemical composition and free radical-scavenging, anti-cancer and anti-inflammatory activities of the essential oil from Ocimum kilimandscharicum. Phytomedicine. **25**;**21**(11), 1298–302 (2014)
23. Mondal, S.: Antimicrobial and Immunomodulatory effects of Tulsi (Ocimum sanctum Linn) [thesis]. Faculty of All India Institute of Medical Sciences, New Delhi, India for the award of the degree of Doctor of Philosophy. pp. 1–141 (2010)
24. Abou El-Soud, N.H., Deabes, M., Abou El-Kassem, L., Khalil, M.: Chemical composition and antifungal activity of Ocimum basilicum L. Essential oil. Open Access Maced. J. Med. Sci. **3**(3), 374 (2015)
25. Behbahani, M.: Evaluation of in vitro anticancer activity of Ocimum basilicum, Alhagi maurorum, Calendula officinalis and their parasite Cuscuta campestris. PLoS ONE **9**(12), e116049 (2014)

26. Mohammadi, M., Majd, A., Nejadsattari, T., Hashemi, M.: Antioxidant and anticancer activities of OcimumbasilicumL. cv. Dark Opal (Lamiaceae). Pharmacogn. Commun. **4**, 48–58 (2014)
27. Naidu, J.R., Gunjan, M., Chen, Y., Sasidharan, S.: Evaluation of in vitro cytotoxic activity of Ocimum basilicum and Mentha spicata extracts. Asian J. Pharm. Clin. Res. **9**, 131–134 (2016)
28. Li, H., Ge, Y., Luo, Z., Zhou, Y., Zhang, X., Zhang, J.: Evaluation of the chemical composition, antioxidant and anti-inflammatory activities of distillate and residue fractions of sweet basil essential oil. J. Food Sci. Technol. **54**, 1882–1890 (2017)

Artificial Neural Network Modelling of Reverse Electrodialysis

Shruti Sen, Rudra Rath, Reddi Kamesh, and Anirban Roy

Abstract Reverse electrodialysis (RED) is an emerging technology with the potential to generate energy from salinity power gradients with limited environmental pollution. In a salinity-gradient power generator (SGP-RED), cation and anion exchange membranes are stacked alternately and due to the difference in concentrations of river and seawater, the diffusion of ions generates an electrochemical potential. In this work, a data-driven approach has been adopted for predicting the performance (Power Density) of SGP-RED power generators over different operating conditions and membrane types, using Artificial Neural Network (ANN) models. Experimental data was gathered and mined from 130 research publications from the last ten years to predict the output Power density of the system against a set of nine input parameters, e.g. membrane types, thickness, resistance, current density, perm selectivity, temperature, etc. A simple mathematical model to estimate the Power density is proposed and used to validate the experimental data before using them with ANN. We used ANN models using Bayesian Regularisation as the back-propagation algorithm (ANN-BR) along with a combination of six sets of pair-wise activation functions amongst Tan-sigmoid, Log-Sigmoid and Linear, in hidden and output layers. The ANN-BR model with Tan-Sigmoid activation functions in both hidden and output layers is shown to predict most accurately the Power density of the SGP-RED system.

Keywords Blue energy · Salinity gradient power · Artificial neural network · Power density

S. Sen (✉) · R. Rath · A. Roy
Department of Chemical Engineering Goa, BITS Pilani, Pilani, Goa 403726, India
e-mail: f20202058@goa.bits-pilani.ac.in

R. Kamesh
Process Engineering and Technology Transfer Department, CSIR-Indian Institute of Chemical Technology, Hyderabad 500007, India

S. Saxena et al. (eds.), *Emerging Materials and Technologies in Water Remediation and Sensing*, Lecture Notes in Civil Engineering 439,
https://doi.org/10.1007/978-981-99-6762-9_4

1 Introduction

Global demand for energy is increasing at an unsustainable rate, which has resulted in the increased importance of renewable energy sources. Salinity gradient energy, generated from mixing river and seawater to produce Gibbs energy, is one such energy source. The product of the energy density of river water times the flow rate of river water gives us the potential of Salinity Gradient Power. There are numerous ways to capture SGP, including RED (SGP-RED), PRO (SGP-PRO), vapour pressure difference utilisation, mixing entropy batteries and capacitive mixing [1]. Of these, RED is one of the most promising techniques capable of exploiting SGP as it has the advantage of continuously converting energy into electricity. There are several other advantages of RED power generation, such as unlimited supply (when river and seawater are used), no production of air pollutants and no thermal pollution or radioactive waste. But, relative to other fuels, the salinity gradient energy content of river water is rather poor. As a result, the investment costs for an SGP plant may be high. A lower return on investment is observed since the transportation cost of water to and within the plants can be substantial in comparison with the generation capacity. Using the average value of energy content in river water to estimate the global potential power generation capacity of SGP, Wick and Schmitt arrived at an approximate value of 2.6 TW [2]. Data collected in 2008 reported that 5 TW energy was used to generate 2 TW of electrical energy in the least efficient coal power plants. Hence, it was concluded that the potential of SGP equivalent to 2.6 TW is more than sufficient to meet the current global energy demands [1].

Power Density (PD) is the power generated per unit area membrane and measures how rapidly the salinity gradient energy gets converted into usable work. According to the reviewed literature, accurate PD predictions can provide an idea of how well-optimised the RED system is. Despite considerable progress in research, optimisation of the RED process and accurate prediction of the PD under various operating conditions and membrane properties are still at an early stage with much scope for improvement.

This study is an effort to predict the performance of a RED-SGP power generator using artificial intelligence and machine learning approaches. The Bayesian Regularisation (BR) backpropagation algorithm along with a combination of three activation functions has been used with Artificial Neural Network (ANN) models to predict the power density for the RED system. Hyperparameters were tuned and the optimum number of hidden neurons was used for accurate prediction. For this work, we have used 600 + data points as the training and testing data sets of the ANN. The reason for selecting the BR backpropagation algorithms was that it shows good results for small-sized datasets [3]. The performance of the algorithm was observed by varying the number of hidden layer nodes and a comparative study was carried out.

2 Problem Formulation

2.1 Red Process Description

The RED system (stack) consists of N number of cells, each cell consisting of a Cation Exchange Membrane (CEM), a salt water (sea water) compartment, an Anion Exchange Membrane (AEM) and a fresh water (river water) compartment. The membranes are sandwiched between spacers, and it is observed that the spacer-less RED is more scalable than spacer-filled RED due to low hydraulic loss. Positive ions selectively diffuse through CEMs from the salt-water compartment to the fresh-water compartment developing +ve potential on one side, while −ve ions selectively diffuse through the AEMs on the other side of a cell causing −ve potential. The resulting potential difference across a cell can be calculated using the Nernst equation. The ionic current in the cells is then converted to an external electron current via Redox reactions at the electrodes. If there is an external circuit connecting the end electrodes, the resulting electric current and potential difference lead to power generation. Figure 1 depicts the schematic of a RED stack.

Although RED is a very attractive proposition for power generation using the vast availability of river and seawater resources, the cost of RED membranes is comparatively higher than other types of membranes (e. g. RO). Hence, it is important to be able to accurately predict and optimise the performance of a RED system under various operating conditions. The Power Density (PD) is the output parameter we aim

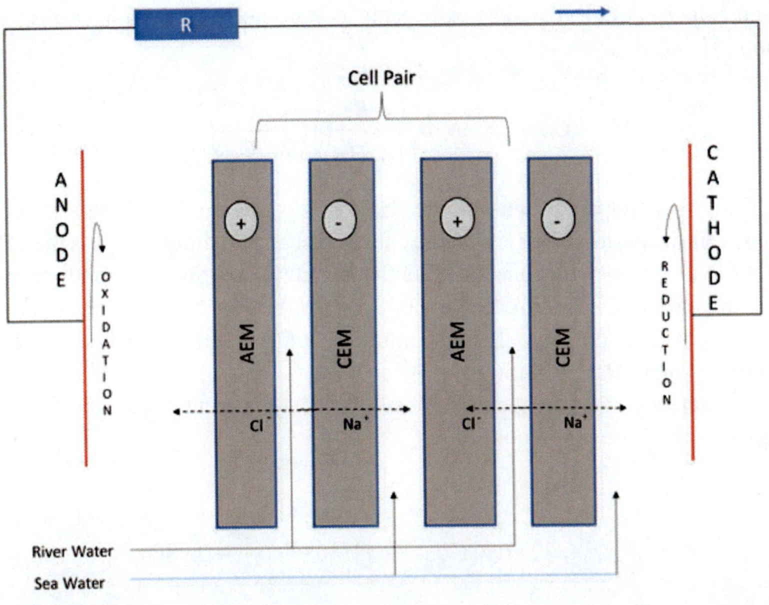

Fig. 1 SGP-RED fuel cell structure

to predict since it directly determines the amount of energy generated by the system. The Power density is dependent on the membrane properties (like perm selectivity), spacer properties, and activity factors of positive ions (Na^+) and negative ions (Cl^-), cell dimensions and number, temperature, internal resistances, current density, flow rate, etc.

2.2 Power Density Model

The RED stack comprises multiple units, with each unit housing a cation exchange membrane (CEM), a section containing a concentrated salt solution (referred to as the 'fuel' mixture), which is a combination of highly concentrated seawater and diluted river water. This unit also contains an anion exchange membrane (AEM) along with a compartment featuring lower salt concentration. The final unit is sealed with an additional cation exchange membrane. The movement of Na^+ ions from the seawater triggers a diffusion process through the cation exchange membranes, inducing a positive potential on the left side of the stack. Similarly, Cl^- ions traverse the anion exchange membrane in the opposite direction, also leading to a positive potential on the left side of the stack. The flow of ions across the membranes takes place when an electrical load is connected to the electrodes. Externally, a standard electric current is observed, whereas internally, there exists an ionic current within the units. This ionic current within the units is transformed into an electron current at the electrodes through redox reactions (refer to Fig. 1). The generated Electromotive Potential (Open Circuit Voltage) over each membrane is derived from the Nernst equation [4] and is given as

$$E_{OCV} = 2N\ \alpha_{CEM} \frac{RT}{zF} ln\left(\frac{a_c.C_c}{a_d.C_d}\right) \tag{1}$$

where E_{OCV} is the Open circuit voltage (EMF), N is the number of cells-membrane pairs, α_{CEM} is the perm-selectivity of the cation exchange membranes, z is the valency (z = 1 for Na^+), R the gas constant, F is the Faraday constant, T is the temperature (in Kelvin scale). The parameters-a_c and a_d-are the ion activity coefficients and the parameters-C_c and C_d-are the concentrations of the concentrated (seawater) and dilute (river water) NaCl solutions.

The ion activity coefficients of NaCl solutions at different concentrations can be estimated by the following equations.

1. For $0 < C_{NaCl} \leq 0.024$

$$a_c = 257.97C_{NaCl}^2 - 11.368C_{NaCl} + 0.9864$$

2. For $0.024 < C_{NaCl} \leq 0.19$

$$a_c = 4.4627C_{NaCl}^2 - 1.6864C_{NaCl} + 0.8947$$

3. For $0.19 < C_{NaCl}$

$$a_c = 0.2461C_{NaCl}^2 - y.3472C_{NaCl} + 0.8947$$

Based on this, a simplified Power Density equation is derived, as the first-principal model:

$$PD(wm^2) = E_{OCV}.I_D - I_D^2.R_s.N \tag{2}$$

where I_D is the current density and R_s is the internal resistance of the RED stack (assuming external load resistance is negligible) and N is the number of cell-membrane pairs.

2.3 *Data Mining*

Data collection was done by doing an extensive literature survey and data mining from existing papers on RED [5–39]. Data points were collected from about 130 research papers and journal articles, on features that affected the power density (PD) of the RED system. The criteria for data mining were divided into two feature categories (i) membrane properties and (ii) process parameters. This resulted in 10 input parameters (features).

Membrane Properties: type of membrane, specific area resistance of AEM and CEM membranes, thickness, active area of the membrane, perm-selectivity.

Process Parameters: number of cell pairs, the temperature of the system, feed solution used, the concentrations of the feed solution, current density.

These input parameters are used to determine the output Power Density (PD) of the RED system. Those papers were considered for data mining which had values for all the above features, and this resulted in the selection of 33 of 130 papers for collecting experimental data. The correctness of the experimental data points was then again validated by calculating the Power Density theoretical values using Eq. (2). The Power densities for various values of current densities derived from the theoretical model are observed to follow a similar trend to that from experimental data, thus confirming the validity of the theoretical Power density model (Fig. 2).

The theoretical model was further used to validate the experimental data to be used with the ANNs. The experimental data were validated against the above first-principal model, retaining those data points that showed a similar trend as from the theoretical model. This resulted in 644 data points that were used for the ANN.

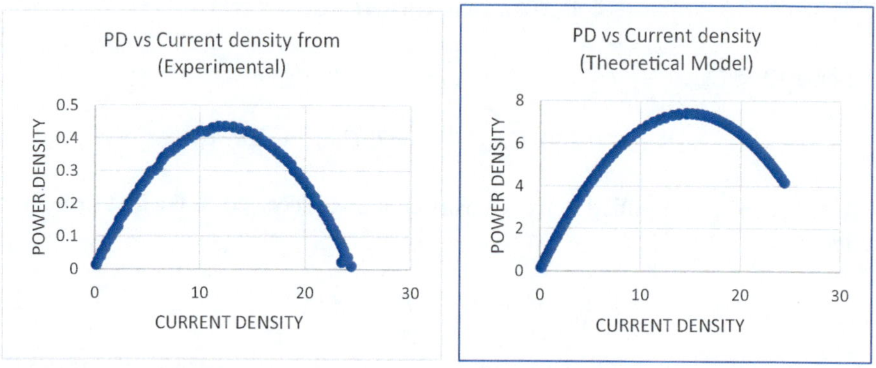

Fig. 2 PD's versus current densities from experimental results and theoretical model

3 Artificial Neural Networks-Based Modelling

In order to accurately predict the Power densities generated by RED systems based on the input parameters, ANN models have been developed with Bayesian Regularisation (BR) backpropagation learning algorithm, with a combination of three different activation functions—TAN-Sigmoid, LOG-Sigmoid and Linear—used in the hidden layer and output layer. The BR backpropagation algorithm enabled fast and stable convergence and also was suitable for training ANNs with small and medium-sized data sets which is the case here. The proposed single hidden layer ANN model is illustrated in Fig. 3.

3.1 Performance Measures for ANN Model

The assessment of ANN model performance and accuracy during both the development (training) and testing phases involves the utilisation of statistical metrics. These metrics encompass the correlation coefficient (R), root mean square error (RMSE), coefficient of determination (R^2) and Mean square error (MSE). RMSE gauges the residual variance, quantifying the model's predictive error in relation to the observed variable. An optimal RMSE value is 0, signifying an ideal fit when close to this value. The correlation coefficient (R) signifies the strength of the relationship between input variables, ranging between 0 and 1. A value nearer to 1 implies a robust relationship, while 0 indicates no correlation between independent variables. A value of 0.5 signifies a moderate correlation. R^2 signifies the extent of variance explicable by the fitted model, with a range of 0 to 1. An ideal R^2 value is 1, indicating a perfect fit.

Our ANN model employs a single hidden layer and determines the optimal number of neurons within it to achieve the best fit. The primary goal is to avert overfitting. The optimal criteria involve achieving a minimum RMSE and an R $\rightarrow$ 1, striking a balance between the two metrics. It's notable that instances exist where both RMSE

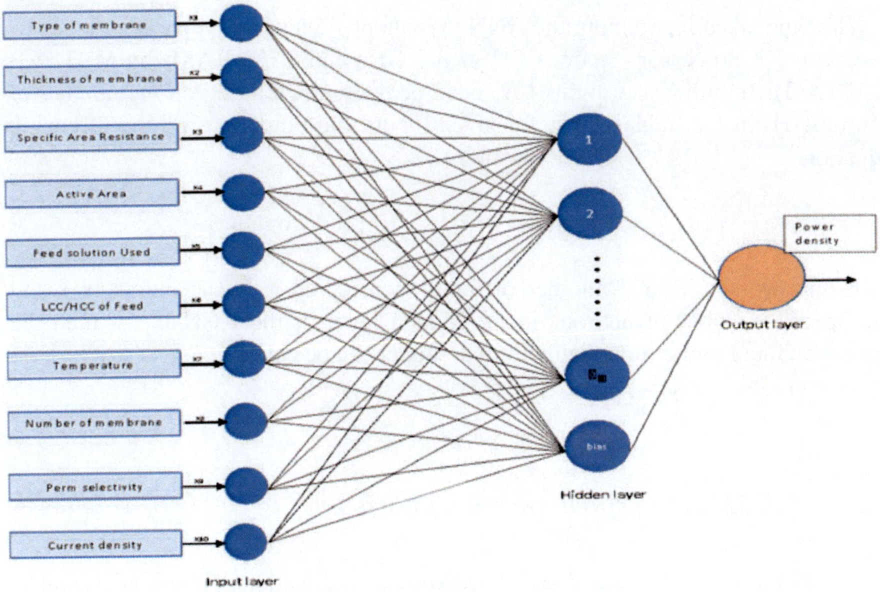

Fig. 3 ANN Model used for SGP-RED

and R can simultaneously be either maximal or minimal. Thus, the selection of hidden layer neurons is guided by meeting these optimal criteria in each scenario. Therefore, the number of neurons in the hidden layer in each case is chosen where the optimal criteria are met. We considered a split of 80–10–10 ratio between training, validation and test data. From the performance of the ANN on the test data (Table 1), the optimal number of hidden nodes is obtained as 15. We've also considered that any further increase in the number of nodes beyond 19 may lead to overfitting.

In the next step, the data inputs to the ANN are scaled in the range of −5 to +5 and the output or target variables are normalised between −1 and +1 to bring all the input and output data to a common magnitude.

Table 1 Finding optimal number of hidden layer neurons

No. of hidden layer neurons	RMSE	R
5	0.1606	0.9992
10	0.3276	0.9991
13	0.3869	0.9931
14	1.8837	0.8698
15	0.0495	0.999
16	0.108	0.9997
17	1.1728	0.9654
18	1.7267	0.9131
19	1.1211	0.9963

The time taken for training the ANN was about 1.5 s using Apple Macbook Air platform, M1 processor (8-core CPU @ 3.2 GHz and 8 GB RAM) by MATLAB R2020b. The number of weights (N_w) can be determined based on the number of neurons (H) in the hidden layer for given inputs and outputs using the following equation:

$$N_w = (\text{input} + 1) \times \text{H} + (\text{H} + 1) \times \text{output}$$

In this work, the total number of inputs considered is 9, the output is 1 and the optimal number of neurons in the hidden layer of the ANN is 15, then the corresponding number of weights for that model can be calculated as $N_w = (9 + 1) \times 15 + (15 + 1) \times 1 = 166$.

3.2 ANN Model Performance Comparison

The ANN models were trained and tested with a combination of Bayesian Regularisation learning algorithms with the following 3 sets of activation function combinations in the hidden layer and output layer, respectively-Tan-Sigmoid and Tan-Sigmoid, Tan-Sigmoid and Linear and Log-Sigmoid and Linear. The performance metrics used for comparison are Mean Squared Error (MSE), Root Mean Squared Error (RMSE), Correlation Coefficient (R) and Coefficient of determination (R^2). The number of hidden neurons is considered 15 in all cases to maintain the set of weights constant.

3.3 ANN with Bayesian Regularisation Backpropagation (ANN-BR) Results

Table 2a, b and c depict the performance during training, test and total data, respectively, of the ANN-based prediction of PD with Bayesian Regularisation backpropagation for various combinations of the activation functions—(a) Tan-Sigmoid + Tan-Sigmoid, (b) Tan-Sigmoid + Linear, (c) Log-Sigmoid + Linear. Although all 4 statistical parameters were evaluated (R, R^2, MSE and RMSE), however, the rankings of model performance were based on R^2 and RMSE values (Figs. 4 and 5).

With R^2 of 0.999 and RMSE value of 0.0336, it can be observed that the ANN-BR model with Tan-Sigmoid in both layers outperforms all other ANN models for PD prediction, closely followed by ANN-BR with Log-Sigmoid + Linear activation combination.

Another backpropagation algorithm is known to perform well on small data sets—Levenberg Marquardt (ANN-LM)—with the combination of the same three activation functions, was also used to predict RED system power densities and compare its

Table 2 **a** Training results **b** Test results **c** Total data results

Activation functions combination	MSE	RMSE	R	R^2
(a)				
Tan-sigmoid + tan-sigmoid	0.001	0.0317	1	0.999
Tan-sigmoid + linear	0.0007	0.0262	1	1
Log-sigmoid + linear	0.0008	0.0275	1	1
(b)				
Tan-sigmoid + tan-sigmoid	0.0025	0.0495	0.999	0.9979
Tan-sigmoid + linear	0.3758	0.6131	0.8308	0.487
Log-sigmoid + linear	0.0092	0.096	0.994	0.9872
(c)				
Tan-sigmoid + tan-sigmoid	0.0011	0.0336	1	0.999
Tan-sigmoid + linear	0.0333	0.1825	0.9989	0.9979
Log-sigmoid + linear	0.0015	0.0386	1	0.999

performance with ANN-BR. In the case of ANN-LM, it is observed that the combination of Log-Sigmoid and Linear activation functions in the hidden and output layers respectively, produced the best prediction performance. Considering both types of learning algorithms—ANN-BR and ANN-LM—it was observed that ANN-BR Tan-Sigmoid + Tan-Sigmoid produced the overall best prediction performance, followed by ANN-BR Log-Sigmoid + Linear, ANN-LM Log-Sigmoid + Linear, ANN-LM Tan-Sigmoid + Linear, ANN-BR Tan-Sigmoid + Linear, ANN-LM Tan- Sigmoid + Tan-Sigmoid, in that order.

4 Conclusion and Future Directions

Modelling and prediction of the performance of a RED stack is a highly non-linear problem. Accurate power density prediction requires efficiently and robustly capturing all the different input parameters—both membrane properties and various operating conditions that the membranes are subjected to. The data-driven ANN framework described in this paper provides a direction to accurately estimate the performance of a RED system to be able to tune various operating parameters for improving the system's performance.

The net power produced by a RED system is dependent not just on the electrical power but is also limited by various losses (hydrodynamic) occurring in the stack. For example, spacers that help maintain the membrane position play an important role by inducing both electrical resistance and thermodynamic resistance. High flow rates, for example, lead to increased power density but reduce overall thermodynamic efficiency. In order to maximise both the power generation and the efficiency of the system, we would need to have a combination of various system design parameters

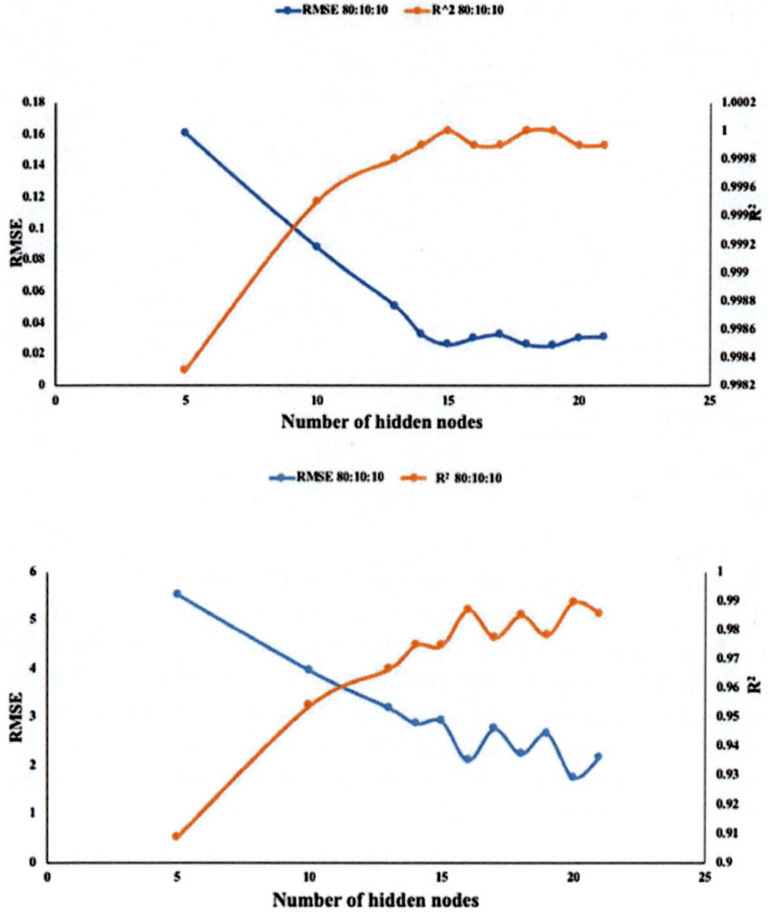

Fig. 4 Tan-sigmoid + tan-sigmoid

like low electrical resistance (thin compartments and membranes), low water friction, high permselective membranes, efficient electrode system, among others [39]. Membranes suitable for RED systems combining low resistance and high permselectivity are not easily available. Development of such membranes at economically viable price points is a challenge. Developing a complex systems model for RED to optimise the performance of the system based on tuning of various operational and design parameters will be the topic of future research.

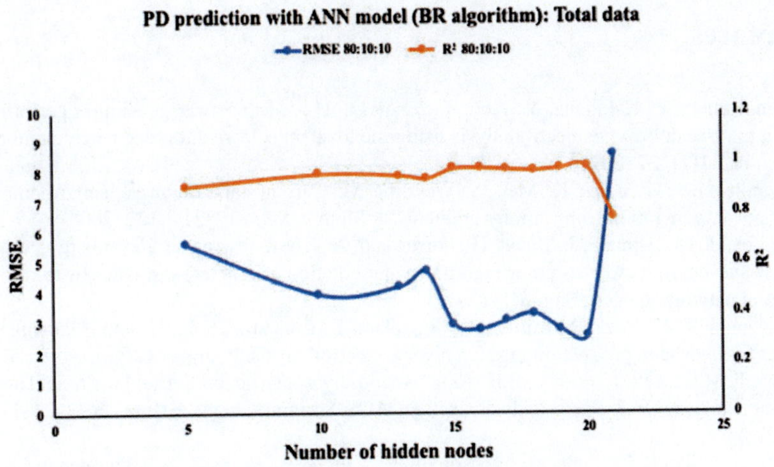

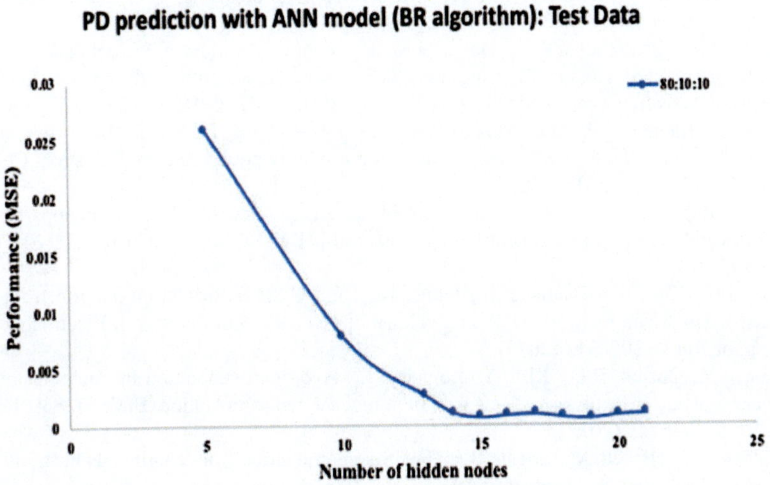

Fig. 5 Tan-sigmoid + tan-sigmoid

Acknowledgements I would like to thank Prof. Anirban Roy for guiding us throughout and Rudra Rath for helping me in every step of the way. This project would not have been possible without their help.

References

1. Mehdizadeh, S., Kakihana, Y., Abo, T., Yuan, Q., Higa, M.: Power generation performance of a pilot-scale reverse electrodialysis using monovalent selective ion-exchange membranes. Membr. **11**(1), 27 (2021)
2. Długołęcki, P., Nymeijer, K., Metz, S., Wessling, M.: Current status of ion exchange membranes for power generation from salinity gradients. J. Membr. Sci. **319**, 214–222 (2008)
3. Garoosiha, H., Ahmadi, J., Bayat, H., Formisano, A.: The assessment of levenberg–marquardt and bayesian framework training algorithm for prediction of concrete shrinkage by the artificial neural network. Cogent Eng. **6**(1) (2019)
4. Krakhella, K.W., Bock, R., Burheim, O.S., Seland, F., Einarsrud, K.E.: Heat to H2: using waste heat for hydrogen production through reverse electrodialysis. Energies **12**(18), 3428 (2019)
5. Hatzell, M.C., Zhu, X., Logan, B.E.: Simultaneous hydrogen generation and waste acid neutralization in a reverse electrodialysis system. ACS Sustain. Chem. & Eng. **2**(9), 2211–2216 (2014)
6. Tufa, R.A., Hnát, J., Němeček, M., Kodým, R., Curcio, E., Bouzek, K.: Hydrogen production from industrial wastewaters: an integrated reverse electrodialysis-water electrolysis energy system. J. Clean. Prod. **203**, 418–426 (2018)
7. Skilbred, E.S., Krakhella, K.W., Haga, I.J.M., Pharoah, J.G., Hillestad, M., del Alamo Serrano, G., Burheim, O.S.: Heat to h2: using waste heat to set up concentration differences for reverse electrodialysis hydrogen production. ECS Trans. **85**(13), 147 (2018)
8. Chen, X., Jiang, C., Shehzad, M.A., Wang, Y., Feng, H., Yang, Z., Xu, T.: Water-dissociation-assisted electrolysis for hydrogen production in a salinity power cell. ACS Sustain. Chem. & Eng. **7**(15), 13023–13030 (2019)
9. Liu, B., Zhang, L., Xiong, W., Ma, M.: Cobalt-nanocrystal-assembled hollow nanoparticles for electrocatalytic hydrogen generation from neutral-pH water. Angew. Chem. **128**(23), 6837–6841 (2016)
10. Han, J.H., Kim, H., Hwang, K.S., Jeong, N., Kim, C.S.: Hydrogen production from water electrolysis driven by high membrane voltage of reverse electrodialysis. J. Electrochem. Sci. Technol. **10**(3), 302–312 (2019)
11. Nam, J.Y., Cusick, R.D., Kim, Y., Logan, B.E.: Hydrogen generation in microbial reverse-electrodialysis electrolysis cells using a heat-regenerated salt solution. Environ. Sci. Technol. **46**(9), 5240–5246 (2012)
12. Watson, V.J., Hatzell, M., Logan, B.E.: Hydrogen production from continuous flow, microbial reverse-electrodialysis electrolysis cells treating fermentation wastewater. Biores. Technol. **195**, 51–56 (2015)
13. Luo, X., Nam, J.Y., Zhang, F., Zhang, X., Liang, P., Huang, X., Logan, B.E.: Optimization of membrane stack configuration for efficient hydrogen production in microbial reverse-electrodialysis electrolysis cells coupled with thermolytic solutions. Biores. Technol. **140**, 399–405 (2013)
14. Zhu, X., Kim, T., Rahimi, M., Gorski, C.A., Logan, B.E.: Integrating reverse-electrodialysis stacks with flow batteries for improved energy recovery from salinity gradients and energy storage. Chemsuschem **10**(4), 797–803 (2017)
15. Kingsbury, R.S., Chu, K., Coronell, O.: Energy storage by reversible electrodialysis: the concentration battery. J. Membr. Sci. **495**, 502–516 (2015)
16. Kim, D.H., Park, B.H., Kwon, K., Li, L., Kim, D.: Modeling of power generation with thermolytic reverse electrodialysis for low-grade waste heat recovery. Appl. Energy **189**, 201–210 (2017)
17. Palenzuela, P., Micari, M., Ortega-Delgado, B., Giacalone, F., Zaragoza, G., Alarcón-Padilla, D.C., Micale, G.: Performance analysis of a RED-MED salinity gradient heat engine. Energ. **11**(12), 3385 (2018)
18. Olkis, C., Santori, G., Brandani, S.: An adsorption reverse electrodialysis system for the generation of electricity from low-grade heat. Appl. Energy **231**, 222–234 (2018)

19. Stone, M.L., Rae, C., Stewart, F.F., Wilson, A.D.: Switchable polarity solvents as draw solutes for forward osmosis. Desalination **312**, 124–129 (2013)
20. Mercer, E., Davey, C.J., Azzini, D., Eusebi, A.L., Tierney, R., Williams, L., McAdam, E.J.: Hybrid membrane distillation reverse electrodialysis configuration for water and energy recovery from human urine: an opportunity for off-grid decentralised sanitation. J. Membr. Sci. **584**, 343–352 (2019)
21. Long, R., Li, B., Liu, Z., Liu, W.: Hybrid membrane distillation-reverse electrodialysis electricity generation system to harvest low-grade thermal energy. J. Membr. Sci. **525**, 107–115 (2017)
22. Micari, M., Cipollina, A., Giacalone, F., Kosmadakis, G., Papapetrou, M., Zaragoza, G., Tamburini, A.: Towards the first proof of the concept of a reverse electroDialysis-membrane distillation heat engine. Desalination. **453**, 77–88 (2019)
23. Zhu, X., He, W., Logan, B.E.: Influence of solution concentration and salt types on the performance of reverse electrodialysis cells. J. Membr. Sci. **494**, 154–160 (2015)
24. Geise, G.M., Cassady, H.J., Paul, D.R., Logan, B.E., Hickner, M.A.: Specific ion effects on membrane potential and the permselectivity of ion exchange membranes. Phys. Chem. Chem. Phys. **16**(39), 21673–21681 (2014)
25. Li, W., Krantz, W.B., Cornelissen, E.R., Post, J.W., Verliefde, A.R., Tang, C.Y.: A novel hybrid process of reverse electrodialysis and reverse osmosis for low energy seawater desalination and brine management. Appl. Energy **104**, 592–602 (2013)
26. Tedesco, M., Cipollina, A., Tamburini, A., van Baak, W., Micale, G.: Modelling the reverse electroDialysis process with seawater and concentrated brines. Desalin. Water Treat. **49**(1–3), 404–424 (2012)
27. Brauns, E.: Towards a worldwide sustainable and simultaneous large-scale production of renewable energy and potable water through salinity gradient power by combining reversed electrodialysis and solar power? Desalination **219**(1–3), 312–323 (2008)
28. Hall, A.N.: Outline of a new thermodynamic model of energetic fuel-coolant interactions. Nucl. Eng. Des. **109**(3), 407–415 (1988)
29. Kwon, K., Han, J., Park, B.H., Shin, Y., Kim, D.: Brine recovery using reverse electrodialysis in membrane-based desalination processes. Desalination **362**, 1–10 (2015)
30. Vermaas, D.A., Saakes, M., Nijmeijer, K.: Power generation using profiled membranes in reverse electrodialysis. J. Membr. Sci. **385**, 234–242 (2011)
31. Farrell, E., Hassan, M.I., Tufa, R.A., Tuomiranta, A., Avci, A.H., Politano, A., Arafat, H.A.: Reverse electrodialysis powered greenhouse concept for water-and energy-self-sufficient agriculture. Appl. Energy, **187**, 390–409 (2017)
32. Mehdizadeh, S., Yasukawa, M., Kuno, M., Kawabata, Y., Higa, M.: Evaluation of energy harvesting from discharged solutions in a salt production plant by reverse electrodialysis (RED). Desalination **467**, 95–102 (2019)
33. Zhou, Y., Zhao, K., Hu, C., Liu, H., Wang, Y., Qu, J.: Electrochemical oxidation of ammonia accompanied with electricity generation based on reverse electrodialysis. Electrochim. Acta **269**, 128–135 (2018)
34. Nazif, A., Karkhanechi, H., Saljoughi, E., Mousavi, S.M., Matsuyama, H.: Recent progress in membrane development, affecting parameters, and applications of reverse electrodialysis: a review. J. Water Process. Eng. **47**, 102706 (2022)
35. Kwon, K., Park, B.H., Kim, D.H., Kim, D.: Comparison of spacer-less and spacer-filled reverse electrodialysis. JJ. Renew. Sustain. Energy **9**(4), 044502 (2017)
36. Pintossi, D., Simões, C., Saakes, M., Borneman, Z., Nijmeijer, K.: Predicting reverse electrodialysis performance in the presence of divalent ions for renewable energy generation. Energy Convers. Manage. **243**, 114369 (2021)
37. Güler, E., Elizen, R., Vermaas, D.A., Saakes, M., Nijmeijer, K.: Performance-determining membrane properties in reverse electrodialysis. J. Membr. Sci. **446**, 266–276 (2013)
38. Altıok, E., Kaya, T.Z., Güler, E., Kabay, N., Bryjak, M.: Performance of reverse electrodialysis system for salinity gradient energy generation by using a commercial ion exchange membrane pair with homogeneous bulk structure. Water **13**(6), 814 (2021)

39. Ju, J., Choi, Y., Lee, S., Kim, H., Jung, N.: Effect of design and operating parameters on power generation in reverse electrodialysis (RED): experimental analysis and modeling. Desalination Water Treat. **191**, 29–39 (2020)
40. Rath, R., Dutta, D., Kamesh, R., Sharqawy, M.H., Moulik, S., Roy, A.: Rational design of high power density "blue energy harvester" pressure retarded osmosis (PRO) membranes using artificial intelligence-based modeling and optimization. Energy Convers. Manage. **253**, 115160 (2022)

Development of Dairy Wastewater Treatment for Valuable Applications

Swati Sucharita Satpathy, Priti Chhanda Ojha, and Debabrata Pradhan

Abstract With the increasing demand for different dairy products, dairy industries have increased their production resulting in the generation of an enlarged volume of waste. Wastewater (WW) is their major waste which contains high concentration of biological and chemical oxygen demand, and different organic and inorganic debris. There is little scope for processing entire WW in terms of overall production cost. Therefore, a major portion of the WW is discarded either after the primary treatment or directly without any treatment in different developing countries. However, different research groups are trying their best to find out a low-cost and eco-friendly solution for developing a suitable treatment process. Current review summarises the recent development in the treatment of dairy WW (DWW) using different techniques. Since the DWW contains a high concentration of organic pollutants, it can be used as a carbon source for the production of bio-energy like methane gas and bioelectricity. For this purpose, microbial fuel cell (MFC) is one of the advanced techniques. Many researchers are working on the development of a suitable MFC for the DWW treatment coupled with bio-energy generation. Further, the MFCs are hybridised to produce high calorific hydrogen gas which will be going to become the common fuel for all types of energy purposes. Direct anaerobic treatment of the WW is another technique used for the production of biogas; however, the longevity of anaerobic reactor has posed many questions. The high organic pollutant containing DWW is used as a substrate for the production of bio-surfactants, bio-plastics and bio-molecules through different biotechnological methods. Although there is potential in the valorisation of DWW streams to produce bio-energy and bio-products while also resolving the waste handling challenges, more work remains to be done to enhance and demonstrate the industrial viability of these technologies.

S. S. Satpathy · P. C. Ojha · D. Pradhan (✉)
Biofuels and Bioprocessing Research Center, ITER, Siksha 'O' Anusandhan (Deemed to Be University), Bhubaneswar 751030, India
e-mail: debabratapradhan@soa.ac.in

P. C. Ojha
Vasudev Higher Secondary School, Talcher, Angul 759100, India

S. Saxena et al. (eds.), *Emerging Materials and Technologies in Water Remediation and Sensing*, Lecture Notes in Civil Engineering 439,
https://doi.org/10.1007/978-981-99-6762-9_5

Keywords Dairy wastewater · Treatment technique · Bio-energy · Bioproducts · Industrial viability

1 Introduction

The continued consumption of finite resources is putting an increasing strain on our loving Earth. Since the beginning of industrial revolution, this trend has only increased. Different minerals and hydrocarbons, examples of the earth's finite natural capital, have been mined and refined to create an ever expanding range of goods and services. Already difficult problem of meeting the growing population's increased demand for water, energy and food takes on a new dimension of complexity. Being a dietary staple dairy products play an important role in ensuring the people's continued access to the nutritious food. With the increasing demand for different dairy products, dairy industries have increased their production. Further, the current techniques applied for the dairy productions pose significant ecological risks due to generation of enlarged volume of different wastes. Dairy wastes include methane, manure and WW [1].

The dairy industry relies heavily on water, making it difficult to dispose the copious amount of WW produced. It is obvious for the DWW to have a high biological oxygen demand (BOD) and chemical oxygen demand (COD) due to the presence of different water pollutants like suspended and dissolved solids, trace and soluble organic components, fats, nutrients, lactose, sulphates and chlorides. In order to make different industrial products or generate energy, these remnants of DWW can be used as raw materials. Pollutants like lactose, fat, minerals and whey protein are just some of the dissolved organic components that found abundance in the DWW. Further, variation in products and processing methods account for the wide range of raw DWW's chemical make-up. Along with the washing detergents and sanitisers, this effluent is highly concentrated in milk components such as casein, inorganic salts and lactose. Casein, fat, lactose, nitrogen, phosphorous, potash and organic matters are all beneficial components of the DWW. Whey produced during the making of cheese and casein is another main content of the DWW. The milky whey contains huge amount of protein, fat, sugar, water-soluble vitamins and minerals. The lactose content of total whey solids is around 75%. Another content of the DWW is the clarified butter sediment waste that produced during the production of clarified ghee or butter [2]. Therefore, the DWW must be effectively treated before environmental discharge to avoid problems like eutrophication, loss of dissolved oxygen, toxicity and biological imbalance in the aquatic habitats. Different eco-friendly treatment techniques for the DWW management are gaining popularity worldwide [3].

Water pollution prevention and the safeguarding of freshwater resources remain a pressing concern in many nations. Effluents from homes and industries are the major contributors to the pollution of different water resources, which drives up the price of water treatment. Additionally, it exposes the water resources to a variety of chemical contaminants and hazardous wastes, which are challenging to remove and treat. For instance, it is very difficult to remediate substances like trichloroethylene, tetrachloroethylene, cyclohexane substituted benzene or toluene, diclofenac sodium, heavy metals, etc. MFC is a new technique beneficial to WW treatment and simultaneous electrical energy generation by using electron-secreting microorganisms, preventing water pollution and the spread of dangerous diseases. It is easy to use, inexpensive and very effective in turning garbage into electrical energy [4, 5].

2 Wastes from Dairy Industry

While providing a varied selection of dairy products, the dairy industry produces an abundance of waste and by-products. Alternative biotechnological uses of these wastes to aerobic and anaerobic treatment systems must be established due to their high COD levels and rich nutritional (lactose, protein and fat) contents. With the advanced technology and widespread products like cheese, yoghurt, butter, cream and ice cream, the dairy industry occupies a prominent position among the food industries. Problems on the farm (such as spillage), in the laboratory (such as improper storage), during transportation (such as damaged packaging) and in the grocery store (such as subpar products) contribute to the massive amounts of milk and other processed dairy products that end up in the trash. Additionally, dairy industry wastes, such as cheese whey, spilled milk and curd chunks, also contribute to the water pollution because they contain different milk residues. Different dairy wastes produced from major dairy products are listed in Table 1. The release of an excessive amount of whey during the cheese production to food waste is one such issue. The high-fat content of many other types of waste and by-products also make their treatment a challenge. The presence of lactose and milk proteins is responsible for the high COD and BOD levels in the wastes. Organic dairy wastes provide a serious environmental concern because of the way they are handled. Whey and other dairy products are challenging to treat in both aerobic and anaerobic systems due to their low bicarbonate alkalinity, high COD concentration and propensity to rapidly acidify. The creation of new bio-based products might utilise dairy industry by-products and wastes as a source of raw materials, which would help to reduce the overall carbon footprint. Once introduced to the bio-refinery, these wastes will aid in the separation and recovery of different bio-products, the growth of microorganisms and the development of bio-economic systems. Due to their low-cost and high-content, outdated milk, cheese whey and other by-products of the dairy industries have been considered as viable substrates in different microbial production methods like bio-energy, enzymes, organic acids, biopolymers, biomass, etc. The creation of bioethanol, biogas and biohydrogen are just a few examples of how these by-products can be used to generate cheap and eco-friendly electricity. In the context of sustainable industrialisation, bringing these cutting-edge technologies into mass production could benefit the economy, the environment and the creation of new jobs. Together, metabolite production and waste utilisation from microbial processes move us closer to our goal of reducing the water

Table 1 Various waste generated from major dairy products [2, 6, 8, 9]

Sl. no.	Major dairy products	Wastage generated from dairy product
1.	Raw milk	Skimmed milk, spoiled milk, spilled milk contain high content of nitrogen and phosphorous
2.	Sterilised and pasteurised milk	Wastewater generated during production, and packaging of milk products contains suspended solids and organic matter
3.	Cheese	Cheese whey waste and whey permeates
4.	Butter, clarified ghee and ice cream	Clarified butter sediment waste generated during the production of butter or clarified ghee
5.	Curd	Curd chunks waste
6.	Yoghurt	Fatty effluents like grease and oil
7.	Milk powder	Sludge produced during solids separation from WW

pollution. The goal of zero-waste can also be aided by the valorisation of these wastes through bioprocesses [6].

Depending on the seasonal change and weather patterns, there are different risks to the environment and public health associated with the direct dumping of dairy wastes. Once the curd separated from the milk, the remaining liquid fraction of the milk is a transparent yellowish-green colour. The formation of dairy sludge, which is strongly scented by butyric acid, is a by-product of casein degradation. The dairy sludge is another damaging dairy waste despite having high levels of decomposable carbon and low heavy metal content. The dairy sludge accounts for between 5 and 25% of the total volume of WW discharged by the dairy industries. The WW treatment process and environmental factors determine its exact composition [7].

3 Impact of DWW on Environment

Milk is composed of approximately 87% water and 13% solids [7]. Dairy wastes can be divided into two general categories, i.e. dairy effluent (DWW) and solid waste. Salts, lipids, chlorides, sulphates, dissolved solids, soluble proteins, D-lactose and chlorides are all components of DWW. Every step of the production and processing of different dairy products, starting from the milk harvesting to the packing of the finished goods, contributes to the generation of wastes. Since dairy products are consumed by humans, it is necessary for dairy processing facilities to be clean. This entails the use of cleaning chemicals, which ultimately end up in waste effluents and solid wastes. Waste raw milk contains a variety of organic and inorganic functional groups of nitrogen, including NO_2^-, NO_3^- and NH_4^+, which are broken down into nitrite. The nitrile contaminates the groundwater and causes methemoglobinemia. The synthesis of numerous valuable products through microbial treatments is made possible by the DWW's high organic matter content, which is favourable for the sustenance of microbial colonies [8, 9]. Composition of different DWW and their effect on environment are listed in Table 2.

Dairy industry uses a lot of water, so it also produces a lot of DWW. White in colour with a pungent odour and a muddy consistency of the DWW are not pleasant to be around. Every litre of processed milk results in approximately 2.5 L of DWW. Production processes, whey processing, and cleaning of containers and piping all contribute to the creation of the DWW. The use of caustic soda for cleaning increases the sodium content. Fats, high-nutrient compounds and detergents are found in non-toxic DWW. For this reason, it is distinguished from the other types of dairy waste by its higher than average temperature (30–40 °C) as well as its high concentrations of COD and BOD_5. Casein breakdown produces dairy sludge, which smells strongly of butyric acid. The dairy sludge accounts for between 5 and 25% of the total volume of treated DWW released by the dairy industries. Its make-up is dynamic, changing in response to factors like WW treatment and the natural setting. Heavy metals like copper and zinc are more concentrated in dairy sludge than in other types of dairy wastes. The high nitrogen content of the bound proteins also contributes to low carbon content [7]. As a result, it poses a threat to different aquatic organisms if it is discharged into the environment without first being treated to remove excess nutrients. Discharging fatty effluents like grease and oil into water bodies causes a thin layer to form on the water's surface, which prevents oxygen from reaching the water below. Dissolved solids, soluble proteins, D-lactose, salts, lipids, chlorides and sulphates are the building blocks of dairy effluents. Casein, lactose, nitrates, phosphates and potassium are also present in DWW [8]. The organic load of the DWW is primarily driven by the milk's carbohydrate, protein and fat content. Problems, especially those related to the organic load on the municipal sewage system, could emerge from the waste streams of dairy industries. The dairy industry is a major source of pollution, which has negative effects on the quality of our air, soil and water. It contributes to the climatic shifts. The DWW quickly degrades in different water reservoirs due to the presence of high organic load, leading to depletion of the dissolved oxygen level. Insects that breed in these conditions spread deadly diseases like malaria, dengue, yellow fever and chicken guinea famine (mosquitoes and flies). Dairy wastes in large quantity are said to be harmful to certain fishes and algae species.

Table 2 Composition of different DWWs and their effects on environment [2, 8]

Sl. No.	Content of DWW	Effects
1.	Raw milk waste	Methemoglobinemia
2.	Sodium chloride residue	Groundwater contamination
3.	Organic matters and suspended solids	High COD in water
4.	By products	High BOD in water
5.	Oils and greases	Eutrophication
6.	Cleaning products residue	Toxic effects on aquatic animals
7.	Nitrogen, phosphorous and ammonia	Decreased crop yield

Algae and bacteria thrive in the dairy effluent, consuming the river's oxygen supply and killing aquatic life. The death of fish is drawn out and painfully slows down due to this process. High levels of total suspended solids can cause issues for aquatic life. If there are particles suspended in the water, the fish may have trouble in breathing because their gills can't get enough oxygen. Flavourings and gelatinous milk are the main contributors to the suspended solids. Milk whey is an extremely polluting by-product due to the high levels of BOD, i.e. 40,000–48,000 mg/L and COD, i.e. 89,000–95,000 mg/L as well as the large amount of phosphorus and nitrogen [2].

4 Microbes Present in DWW

Microorganisms are able to grow using the DWW as a growth substrate. Bacteria, fungi and protozoa are just some of the organisms that found in the untreated dairy wastes (Table 3). *Sphaerotilus natans* and other sewage fungi thrive in the DWW. The DWW has also been used to isolate yeasts like *Saccharomyces* sp., *Candida* sp. and *Cryptococcus* sp. The fungal strain, i.e. *Aspergillus aculeatus* was originally discovered from the dairy sludge. Different bacteria from the lactic acid bacteria family tend to do well in the dairy waste and this was reflected in the microbial composition of the raw dairy waste. Also the dairy sludge releases airborne microorganisms that can able to live and reproduce. Different types of heterotrophic microorganisms are found in discarded dairy products. Gram-negative bacteria accounted for 30% of the microbes population found in the dairy waste. They mainly take the forms of rods, cocci and coccobacilli. *Pseudomonas aeruginosa, Enterobacter, Pseudomonas fluorescens, Proteus* sp., *Escherichia coli, Alcaligenes* sp., *Citrobacter* sp., *Pseudomonas nitroreducens, Klebsiella pneumoniae* and *Acinetobacter baumannii* were among the bacterial species isolated from different dairy wastes. *Micrococcus luteus, Staphylococcus* sp., *Bacillus subtilis, Bacillus cereus, Sporolactobacillus* sp. and *Streptococcus faecalis* were among the Gram-positive strains isolated from the dairy wastes and categorised by coccobacilli shape. Seasonal and other environmental changes affect their types and abundance of microorganisms found in the dairy wastes [7]. McAteer et al. (2020) demonstrated how the reactor configuration influences the microbial presence during an anaerobic treatment of a DWW sample [10]. The dominant genus discovered during the low-temperature anaerobic treatment was *Lactococcus.* The lactic acid bacteria known as *Lactococcus* quickly converted proteins and carbohydrates to lactic acid. High abundance of *Methanobacterium* was found in the community. *Methanosaeta* was identified as predominant active methanogen throughout the experiment. *Smithella* and *Actinomycetaceae* were also found. Bacteroidetes which is a hydrolyser and linked with fermentative reactions of organic substrates was also present throughout the experiment. Acidogens responsible for volatile fatty acids production represented by *Trichococcus* and *Actinomycetaceae* were more relatively abundant at low temperature [10].

Table 3 Major microorganisms present in different dairy wastes

DWW source	Major microorganisms present	Reference
Cheese whey powder solution from the Industry Cargill Agri'cola , Brazil	*Thermoanaerobacterium*, *Thermohydrogenium kirishiense* and *Thermoanaerobacterium thermosaccharolyticuum*	[41]
Cheese whey from a dairy factory, Rafaela city, Santa Fe, Argentina	*Escherichia coli* and *Pseudomona aeruginosa*	[42]
Whey WW	*Kluyveromyces marxianus* and *Saccharomyces cerevisiae*	[43]
Synthetic whey WW	*Vanrija albida*	[8]
Deproteinated cheese whey	*Actinobacillus succinogenes*	[9]
Cheese whey from a cheese factory (INNOLACT S.L.) located in Galicia, Spain	*Proteobacteria*, *Firmicutes*, *Bacteroidetes* and *Nitrospirae*	[44]
DWW	*Rhodopseudomonas rutila* and *Bacillus cereus*	[9]
Whey WW	*Aspergillus niger*	[6]
Cheese whey WW	*Leuconostoc mesenteroides*, *Xanthomonas campestris*, *Komagataeibater xylinus* and *Clostridium* spp.	[46]
DWW	*Kluyveromyces marxianus, Debaryomyces hansenii* and *Galactomyces geotrichum*	[6]

5 Valuable Bio-products from DWW

By using various biological processes like microbial fermentation and anaerobic digestion, the DWW can be cleaned up and converted into a number of useful bio-products. As discussed before, the DWW are a great source of lactose and other nutrients that encourage microbial development for the creation of bio-products, despite the fact that they are polluting [6]. Different valuable bio-products from the DWW are shown in Table 4. Mozejko-Ciesielska et al. (2022) set up an experiment for valorisation of cheese whey mother liquor and cheese whey for biopolymer (scl-polyhydroxyalkanoates homopolymers and heteropolymers) production by *Paracoccus homiensis* [11].

Table 4 Different valuable bio-products from DWW [2, 6]

Bio-products produced from dairy waste by-products	
Bio-products	Source by-products
Biofuels (biomethane, bioethanol, biohydrogen and biodiesel)	Cheese whey; whey powder; whey WW
Enzymes	DWW; whey
Solvents (biobutanol)	Whey waste
Vitamins	Whey
Biosurfactant	DWW
Glycerol	Whey waste
Organic acids	Sweet whey; yoghurt production waste
Bioenergy	Fatty waste, dairy waste
Biopolymers	Whey
Hormones	Whey
Ribonucleotides	Whey
Biomass	Whey; dairy effluent
Aroma compounds	Whey waste
Bioplastic	Whey; dairy industry waste
Biofertilisers	Whey waste
Polysaccharides	Deproteinized whey; whey permeate

6 DWW Treatments

The most common methods used to assess the value of everyday trash are biotechnological production, aerobic treatment, anaerobic digestion, bioremediation and bioaugmentation. There are many different kinds of colloidal particles in the DWW. Traditional techniques, such as membrane filtration, ion exchange, coagulation, flocculation, solvent extraction and flotation, can be used to get rid of the colloidal DWW particles. Some traditional techniques and their efficiency have been summarised and are given in Table 5. Combination of flocculation and coagulation uses synthetic coagulants like chitosan, fava bean, *Moringa oleifera* and common bean in order to remove pollutants from the DWW. The DWW can be treated through membrane separation. Proteinaceous materials adhere to the membranes during the membrane separation leading to an unpleasant odour. Adding absorbents and coagulants before membrane separation may be a viable option for resolving the bad odour issue. Reverse osmosis and nanofiltration are two cutting-edge techniques that show promising for recycling water, lactose and proteins. These cutting-edge innovations are superior to older systems in many ways, including being more reliable, adaptable, environmentally friendly and cost-effective. Anaerobic digestion is widely used as an efficient technique to treat the DWW. A total of four processes, i.e. hydrolysis, acidogenesis, acetogenesis and methanogenesis, come into play during anaerobic digestion.

Followed by the anaerobic digestion, the sludge generated in the process is further stabilised. An array of integration technologies has emerged for the treatment of DWW to generate value-added products, complementing the efficacy of anaerobic digestion. Bio-electrochemical systems are novel approaches that merge the electrochemical and biological methods to generate hydrogen and other useful products. Some examples of these interconnected technologies are MFCs, enzymatic biofuel cells, microbial desalination, microbial electrosynthesis and microbial solar cells. It is possible to make a wide range of useful things out of the DWW. Bioplastics, biodiesel, biochemicals, ethanol, biofuel, functional beverages, biohydrogen, biocatalysts, organic fertiliser, single-cell proteins, microbial fuel cells and organic acids are just some of the products that can be made with this versatile waste. Polyhydroxyalkanoates are such products that have found widespread use and can be manufactured from the milk whey of DWW. There are lactose, protein and fat in the milk whey, which can be fermented to polyhydroxyalkanoates. Furthermore, milk whey does not need any pretreatment steps like enzymatic or chemical hydrolysis, which is a huge advantage. *Bacillus megaterium* strain was used in this study to produce polyhydroxyalkanoates from the dairy whey. The research led to significant gains in output. In another investigation, cheese whey was used to first generate carboxylic acid via anaerobic fermentation, and then polyhydroxyalkanoates via aerobic fermentation. Biodiesel can be made from a variety of sources, including dairy wastes. Both chalcone 3-hydroxylase and flavanone 3-hydroxylase can be manufactured from the concentrated whey. The DWW typically has a pH between 4.7 and 11.0, which indicates the range and number of disinfectants needed for sanitisation. The classic approach to lowering the COD levels and producing bio-energy is anaerobic digestion of dairy effluent and solid waste. However, the feed pH is drastically lowered due to the formation of volatile fatty acids (VFA) and other organic acids during the fermentation process of breaking down whey (acid/sweet), dissolved nitrogenous and phosphoric solids, proteins and so on [8]. Different DWW treatment approaches are shown in Fig. 1.

The potential of the alga *Chlorella pyrenoidosa* for the treatment of dairy effluent and the extraction of biofuel was investigated by Kothari R et al. The recognised eutrophication-causing agents nitrate and phosphate levels in water bodies were decreased by 60 and 87% in the influent and by 49 and 83% in the effluent, respectively. Due to their crucial function in carbon dioxide fixation, microalgae have recently gained a lot of interest in the treatment and recycling of WW. For the algal feedstock, WW is regarded as a potentially sustainable growing medium. Large-scale industries that handle agricultural and animal products can frequently be found to be the source of algal biomass brought on by nutrient leakage [12]. Three categories can be used to group dairy waste treatments. The natural process, which involves the direct release of untreated waste into the environment, such as the use of whey as feed, is included in the first category. Anaerobic and aerobic methods, like anaerobic filters and aerobic lagoons, are used in biological treatments. The most effective ways to remove organic material from dairy waste are these treatments. Lastly, milk fat and colloidal proteins in dairy waste can be reduced via physicochemical methods.

Table 5 Traditional techniques and their efficiency for the DWW treatment

Process	Source of DWW	Efficiency	Reference
Ultrafiltration	Local sweet industry, Hindustan Sweets Pvt. Ltd, Jadavpur, India	Lactose recovery—90%; Protein recovery—80%	[38]
Anaerobic co-digestion	Milk whey WW mixed with potato stem	Biogas production—0.349 m^3/kg; Electric power generation—12,800 kW	[39]
Aerobic fermentation	Cheese whey WW	Nisin production—26 TPA; Lactic acid production—282 TPA	[40]
Moving bed biofilm reactor	Synthetic WW mimic dairy WW	COD removal—95%	[34]
Anaerobic fluidized-bed reactor	Cheese whey powder solution from the Industry Cargill Agri'cola, Brazil	Hydrogen yield—3.67 mol H_2/mol lactose	[41]
Horizontal subsurface flow constructed wetland	Cheese whey from a dairy factory, Rafaela city, Santa Fe, Argentina	Removal efficiency: Suspended solids—78.4%; BOD—57.9%; COD—68.7%; Nitrate—47%; Nitrite—98%	[42]
Fed batch fermentation with both aerobic and anaerobic treatment	Synthetic whey	COD removal—95%; Biomass—0.4 g	[8]
Stirred tank anaerobic reactor	Cheese whey from a cheese factory (INNOLACT S.L.) located in Galicia, Spain	Caproic acid yield—0.4 g/L/d	[44]

(continued)

Table 5 (continued)

Process	Source of DWW	Efficiency	Reference
Coagulation and flocculation	Cheese whey WW	COD removal—43%; BOD removal—67%	[45]
Coagulation → Sedimentation → Constructed wetland	DWW from Teagasc Moorepark Dairy farm, Fermoy, Co. Cork, Ireland	≥99% removal of COD, nitrogen, phosphorus, suspended solids and turbidity	[47]

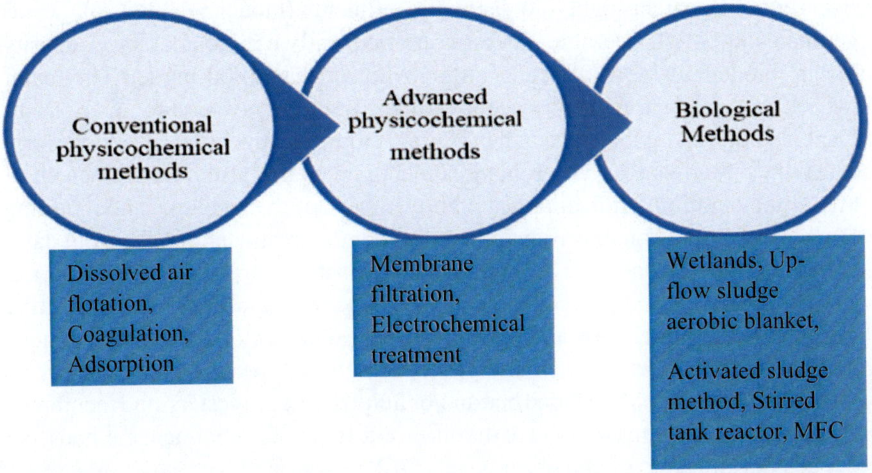

Fig. 1 Different DWW treatment approaches [8, 16]

Physical, chemical and biological processes can help dispose of garbage in accordance with environmental requirements, but they are costly [7]. For the biochemical methane production and biodegradability of DWW (42–57% of COD removal at 72 h of incubation), Ferreira TF et al. used a hybrid treatment that included granular anaerobic sludge as inoculums and whole cells of *Penicillium citrinum* with high lipase activity. Compared to previous pretreatment methods, the hybrid treatment demonstrated a higher rate of methane synthesis. Anaerobic sludge combined with low-cost agro industrial wastes can potentially lower the cost of using microbial cultures to produce substantial amounts of biogas from *P. citrinum* whole cells [13]. Shen et al. (2018) pretreated the dairy manure with ultrasonication for enhanced electricity generation of MFC [14]. Power density obtained with ultrasonic treatment was 241% higher than the untreated. Dairy manure used as main substrate in the dual chambered MFC generated maximum 102 mW/m^2 power density with 102 mA/m2 current density. The COD removal efficiency was 74.6%, and energy of over 10.5 kJ/kg (dry manure) was obtained after 203 d of treatment [14]. Raja et al. (2022) used microalgae *Coelastrella* sp. for WW treatment with electricity production at lab-scale reactor [15]. *Coelastrella* sp. removed 90.38%, 90.24%, 66.75% of NH_4–N, NO_3–N and P, respectively, during 15 d. A removal efficiency of 67.15%, 69.44% and 83.51% were obtained for BOD, COD and total organic carbon, respectively.

7 Traditional DWW Treatments

DWW can be treated using both aerobic and anaerobic techniques to achieve a high level of organic removal efficiency. In contrast to anaerobic processes, which are frequently used to treat highly contaminated effluents (biodegradable COD levels larger than 4000 mg/L); aerobic processes are frequently used to treat lower-density effluents (biodegradable COD concentrations lower than 1000 mg/L). The major drawback of aerobic treatment systems is their high energy requirements. Dairy effluent also offers a good environment for anaerobic treatment because it is warm and heavily contaminated. Anaerobic treatment procedures also have the benefit of not requiring aeration, producing just a little amount of extra sludge, and requiring less space in comparison to aerobic processes. For the treatment of WW from dairy production, numerous aerobic biological treatment methods have been developed, including activated sludge, sequencing batch reactors, traditional or percolating filters, membrane bioreactors and rotating biological contactors. Anaerobic contact process, up-flow anaerobic filter, completely stirred tank reactor, anaerobic digestion, fixed-bed digesters, expanded bed and/or fluidized-bed digesters, and membrane anaerobic reactor system are some of the anaerobic biological treatment methods that have been developed to treat dairy products [16].

Traditional methods (aerobic and anaerobic processes) for the treatment of DWW have a number of drawbacks, including the expense of the land, the climate, the requirement for sludge recycling and more. Although aerobic operations need a lot of energy, anaerobic processes must be integrated with aerobic processes to meet discharge standards. But for the treatment of WW, physicochemical techniques are promising and efficient [17]. Using anaerobic digester systems (biochemical methane potential assay), Johannesson et al. (2020) produced biogas from liquid dairy manure, dissolved air flotation waste and dry poultry manure. A 25% and 161% increase in biogas output from solid poultry manure and liquid dairy manure, respectively, were discovered in co-digestion compared to monosubstrates only. The specific methane yield ranged from 236 to 398 ml CH4g VS-1. [18]. Chu and Wang (2017) used an anaerobic batch fermenter for biohydrogen and biomethane production from dairy cow solid waste. The average hydrogen production rate was 150.59 mL/L/d and the average methane production rate was 395.41 mL/L/d [19]. Using a two-stage digestion in induced bed reactors, Zhong et al. (2015) examined the synthesis of anaerobic hydrogen and methane from dairy waste. Methane production and more COD removal in two-stage digestion process showed greater energy gain. The two-stage digestion process produced 71.7 mL of hydrogen and 61.0 mL of methane overall from 1.0 g of dairy waste. The overall effectiveness of COD elimination was

88.2% [20]. Lhanafi et al. (2018) used batch reactor for co-digestion of dairy waste for enhanced methane production. Maximum methane yield found to be 93.3 L per kilogramme of the dairy waste at pH 8.0 [21].

8 Treatment Using MFC

Today, MFC technology is receiving more attention than ever before because it is widely regarded as a viable option for eliminating bio-waste and producing renewable power. Although MFC has the potential to significantly expand access to clean energy, the fact that its output is not yet sufficient to meet demand remains a major hurdle. The effectiveness of MFC can be improved in a number of ways, including through the selection of suitable microorganisms, the design of the reactor, the use of the right material for the membrane and the selection of the right element for the electrodes. Intensifying processes in the right way could result in the creation of a low-cost, stable power generation tool, in addition to improvements in WW treatment. Most WW contains a lot of toxic organic compounds, which is bad news for the microorganisms in MFCs. Therefore, a biochemical process can be better designed by studying the kinetics of its growth and the rate at which it utilises substrates. Selecting the optimal level of organic compounds to be degraded can be evaluated through kinetic studies because the rate of organic compound removal is indirectly interrelated to electricity generation and bacterial growth. Therefore, to obtain more energy from organic substances present in WW, the selection of proper microorganism having high specific growth rate and specific utilisation rate played a vital role. It is clear that various microorganisms were used as a biocatalyst to generate energy in MFC. These microorganisms include *Geobacteria* spp., *Dessulfobulbus* spp., *Klebsiella* spp., *Rhodoferax* spp., *Rhodopseudomonas* spp., etc. [22]. The DWW treatment using different reactors is listed in Table 6.

With just slight alterations to the MFC architecture and operational circumstances, MFC is a dynamic treatment technique that can use a variety of WW types, in contrast to conventional WW treatment methods. MFC technology offers the ability to reduce carbon emissions from the production of electricity and provide an environmentally benign method for treating a variety of industrial effluents from the petrochemical, tannery, brewery, dairy, textile, pulp/paper and agro-industries [23]. MFCs are discovered to be viable tools to extract bio-energy from WW and simultaneously clean the WW, hence lowering operational expenses associated with WW treatment using traditional techniques [24]. Anode in anodic chamber and cathode in cathodic chamber, separated by a proton exchange membrane, make up a basic MFC configuration. MFC works on the basis that biocatalysts oxidise organic substrates in the anodic chamber, liberating protons and electrons as well as creating CO_2 in the process. Protons are simultaneously moved from the anode chamber to the cathode chamber via proton exchange membrane as the anode collects electrons and transmits them to the cathode through an external circuit. Water is created at the cathode when protons and oxygen combine with electrons [24]. Sekar et al. (2019) used MFC to

Table 6 Different types of reactor used for DWW treatment

WW type	Reactor type	Anode	Cathode	Membrane	Inoculum	Power density	Current density	Open circuit voltage	Pollutant removal efficiency	References
Synthetic medium with progressively increasing DWW	Air-cathode MFC, operating in semi-continuous down-flow mode	Carbon cloth and graphite cylinders	Carbon cloth	Nafion™ 117	*Shewanella oneidensis* and *Clostridium butyricum*	3.5 W/m^3	1.0 A/m^{-3}		COD—89%, BOD—91%, CE—4.5, organic nitrogen—85%, phos-phorus—92%, nitrate—100% and sulphate—88%	[3]
DWW, Gomati Industry Agartala, India	Single-chamber MFC	Carbon cloth	Platinum	Nafion 117	*Shewanella* algae-MTCC-10608 and *Escherichia coli*-K-12-MTCC-1302	62.27 mW/m^2	238 mA/m^2	652 mV	COD—95.31% and CE—31.58%	[5]
DWW	Single-chamber MFC	Carbon cloth	Carbon coated on platinum	Nafion 117	*Shewanella* algae (MTCC-10608)	50 mW/m^2	141 mA/m^2		COD—92.21% and CE—27.45%	[29]

(continued)

Table 6 (continued)

WW type	Reactor type	Anode	Cathode	Membrane	Inoculum	Power density	Current density	Open circuit voltage	Pollutant removal efficiency	References
DWW	Single-chamber MFC	Carbon cloth	Carbon coated on platinum	Nafion 117	*Escherichia coli* -K-12	1.05 W/m^2	8.01 A/m^2	654 mV	COD—95.45% and CE—67.53%	[30]
DWW	Single-chamber MFC	Carbon cloth	Carbon cloth	Nafion 117	*Pseudomonas aeruginosa*-MTCC-7814	105 mW/m^2	313 mA/m^2	1025 mV	COD—95.84% and CE—37.13%	[31]
Synthetic DWW	Dual-chambered MFC	Carbon cloth	Carbon cloth	TiO_2—SPEEK composite	Mixed culture of anaerobic organism	1.22 W/m^2		0.635 V	COD—90%, BOD—87%, total organic carbon—92% and turbidity removal—89%	[32]

(continued)

Table 6 (continued)

WW type	Reactor type	Anode	Cathode	Membrane	Inoculum	Power density	Current density	Open circuit voltage	Pollutant removal efficiency	References
DWW	Air-cathode MFC setup in upflow mode	Carbon cloth	Carbon cloth	Nafion™ 117	*Shewanella oneidensis* and *Clostridium butyricum*	0.48 W/m^2			COD—93%, BOD—95%, nitrate—100%, organic nitrogen—57, sulphate 90% and organic phos-phorus—90%	[33]
Synthetic DWW	Up-flow anaerobic sludge bed reactor (3.5 L)				Anaerobic sludge granules				COD—82.9% and methane—68.9%,	[10]
Synthetic DWW	Moving bed biofilm reactor								COD—95%	[34]
DWW	Dual-chambered MFC	Copper doped iron oxide nanoparti-cles	Carbon papers	Nafion 115 &117, and DuPont™ Nafion	Native microorgan-isms in DWW	161.5 mW/m^2	270 mA/m^2	778 mV	COD—75% and CE—20.9%	[25]

(continued)

Table 6 (continued)

WW type	Reactor type	Anode	Cathode	Membrane	Inoculum	Power density	Current density	Open circuit voltage	Pollutant removal efficiency	References
Acid orange 7 + cheese water and yoghurt water + Acid orange 7	Dual-chambered MFC	Graphite	Graphite	Nafion N966	Anaerobic activated sludge	44.04 mW/ m^2	130.12 mA/ m^2	332.4 mV	COD—84.57%, decolourisation – 92.18% and CE—1.76%	[27]
DWW	Moving bed biofilm reactor				Activated sludge				COD—95%	[34]
DWW	Dual-chambered MFC	Graphite	Platinum coated titanium	Nafion 117	*Pseudomonas aeruginosa, Klebsielaoxytoca* and *Lactobacillus plantarum*		0.026 A		COD—35.39%	[35]
DWW	Single chambered microbial electrolysis cell	Carbon cloth	Carbon cloth		Anaerobic biogas digester slurry	152 mW/ cm^2	19 mA		COD—95% and H_2 generation—32 mL/L/d	[36]
DWW	Dual-chambered MFC	Plain graphite plate	Plain graphite plate	Nafion 117	DWW	2.7W/m^2			COD—91% and CE—17%	[37]

treat real DWW. The performance of MFC exhibited a power density of 161.5 mW/m^2 and COD removal efficiency of 75% [25]. To treat the DWW, Marassi et al. (2022) created an MFC that was injected with a group of *Shewanella oneidensis* and *Clostridium butyricum* bacteria. The findings demonstrated that the air–cathode MFC fed with the DWW may be employed for high-efficiency energy recovery and organic load removal, as well as the removal of acute toxicity. The MFC process showed high efficiency for the removal of nitrate (100%), organic nitrogen (85%), sulphate (88%) and phosphorus (92%) with power density of 3.5 W/m^3 [3]. Sanjay and Udayashankara, (2021) set up a dual chambered membraneless MFC to treat DWW which removed 92.2% of COD, 88.02% of BOD and 76.3% of total dissolved solids. The maximum electricity generated was 644 mV [26].

Dairy effluents can be aerobically digested in a variety of bioreactors, such as up-flow anaerobic sludge blanket reactors, stirred anaerobic sequencing batch reactors, fluidized-bed reactors, plug flow reactors, up-flow fixed-film loop reactors and rotating anaerobic contacts. As a result of their high organic content, dairy industry waste products may be used as a substrate for methane production through anaerobic processes. But because of its high COD, low pH value and lack of alkalinity, anaerobic treatment of whey presents a significant difficulty. However, other dairy wastes (such as milk and yoghurt) with lower COD values may not be suitable for anaerobic digestion in a conventional continuous stirred tank reactor-type biogas plant. According to the data collected from multiple reactors, the up-flow anaerobic sludge blanket reactor is the most effective system for removing COD. Hydrogen gas can be extracted from many types of dairy wastes including sewage sludge, whey, whey powder, and dry milk powder. However, current technology only allows for a maximum of three moles of hydrogen to be extracted from a single unit of lactose consumption, even though theoretically eight moles could be extracted. It is difficult for the hydrogen-producing microorganisms to hydrolyse protein-rich dairy wastes. Various fermentation conditions have been tried to increase the hydrogen yield from dairy wastes, with varying degrees of success. Hydrogen production in a continuous stirred tank reactor is one option at an organic loading rate of 138.6 g lactose/L/d, a molar yield of 2.8 mol H_2/mol lactose was achieved. Dark fermentation in anaerobic conditions is another method for producing hydrogen, and it may be possible to use this technique to make hydrogen on an industrial scale, especially if it is combined with other methods. The first step involves breaking down lactose into its component monomers; the second involves lactic acid bacteria homofermenting glucose and galactose into lactate; and the third involves anaerobic fermentation of lactate to hydrogen and volatile fatty acids. The maximum yield of biohydrogen obtained from a microbial consortium using this technique was 4.13 mol H_2/mol lactose. However, it is crucial to remember that the dark fermentation process is intrinsically complex and depends on a wide range of factors, including the make-up of the feedstock, the pretreatment, the amount of the inoculums, the kind of reactor and the operating circumstances (including temperature, pH, fermentation time, organic loading rate, hydraulic retention time and so on). In addition, during co-fermentation in an enlarged granular sludge bed reactor, the ideal hydrogen yield 120 mmol H_2/g COD was attained using a cheese whey/crude glycerol ratio of 5:1. In this reactor setup, increasing the concentration of crude glycerol repressed H_2 production, which allowed for the production of propionic acid as a by-product. Alternate methods for biohydrogen production from cheese whey in a lab setup involve the use of microbial electrohydrogenesis cells. As a by-product of the dairy industry, dairy waste is a viable material for use in biopolymer manufacturing. The use of enzymatic fuel cells (EFC) to create electricity from whey is an alternative to MFC. This redox-active enzyme-based system is viewed as a cutting-edge method for producing renewable energy. A study employed EFC to convert cheese whey into energy while immobilising cellobiose dehydrogenase. This enzyme is a rare anodic enzyme that produces bioelectrical energy from lactose, cellodextrins, or cellobiose and is renowned for its capacity to transfer electrons. Under ideal circumstances, cheese whey in the EFC showed a maximum power density of 1.839 W/cm^2 [6]. Tajdid et al. (2020) designed an MFC with mixed condition where they replaced the usual nutritional agents with DWW, for the treatment of acid orange 7 WW. The highest rates of COD elimination (84.57%) and decolourisation (92.18%) were achieved [27].

The catalyst- and mediator-free membrane MFC was developed by Mansoorian et al. (2016) to treat DWW and produce bioelectricity at the same time. Power density peaked at 621.13 mW/m, while maximum current density was 3.74 mA. COD, BOD_5, NH_3, NH_4^+, dissolved phosphorus, phosphorus in suspended solids, SO_4^{2-}, total suspended solids and volatile suspended solids all had their maximum removal efficiencies of 90.46%, 73.22%, 31.18%, 72.45%, 39.43%, 70.17% and 64.6%, respectively [28].

9 Conclusions

Difficulties lounge in the dairy industries due to the copious amount of WW they produce. Presence of different water pollutants like suspended and dissolved solids, trace and soluble organic components, fats, nutrients, lactose, sulphates and chlorides make the DWW rich of BOD and COD. They affect our environment in many ways. It is necessary to minimise the pollutants level in the DWW before discarding in to the environment. There are different scopes for minimising the pollutants level in the DWW. One of the scopes is that these remnants of the DWW can be used as raw materials for the production of different valuable products and energy. This scope is further emphasised due to the presence of the diverse microorganisms present in the DWW. The microorganisms have their natural metabolic processes which can be exploited in the utilisation of the pollutant rich DWW for the valuable productions. In this context, there are different physicochemical processes combined with the biotechnology to develop an eco-friendly process at the reasonable cost. Wide range of products like biofuels, enzymes, solvents, vitamins, biosurfactant, glycerol, biopolymers, hormones, ribonucleotides, biomass, aroma compounds, bioplastic, biofertilisers, polysaccharides and bioelectricity are produced from the DWW by using the suitable process. Traditional as well as newly developed techniques, such as membrane filtration, ion exchange, coagulation, flocculation, solvent extraction, flotation, reverse osmosis, nanofiltration, aerobic processes, anaerobic digestion and bio-electrochemical, are used for the above purpose. However, the treatment of DWW has a number of difficulties, including the expense of the land, the climate, the requirement for sludge recycling, non-uniform concentration flow and many more. Further, because of its high COD, low pH value and lack of alkalinity, anaerobic treatment of whey presents a significant difficulty. Every development needs the comprehensiveness. At present the techniques used for the DWW treatment are suitable but not uniform. The processes are pollutant specific. Therefore, there are a lot of scopes for the development of a standard process for the DWW treatment and their utilisation. Synchronisation of biotechnology with the traditional water treatment technology has a lot of scope for the purpose. Government and the industrial partners should come forward to encourage the research related to the development of a comprehensive process to deal with the huge amount of DWW.

Acknowledgements The authors are grateful to Prof. (Dr.) Manojranjan Nayak, President of Siksha 'O' Anusandhan (Deemed to be University) for providing infrastructure and encouragement throughout the research.

References

1. Luqman, M., Al-Ansari, T.: A novel solution towards zero waste in dairy farms: a thermodynamic study of an integrated polygeneration approach. Energy Convers. Manage. **230**, 113753 (2021)
2. Ahmad, T., Aadil, R.M., Ahmed, H., Rahman, U., Soares, B.C., Souza, S.L., Cruz, A.G.: Treatment and utilization of dairy industrial waste: a review. Trends Food Sci. Technol. **88**, 361–372 (2019)
3. Marassi, R.J., López, M.B.G., Queiroz, L.G., Silva, D.C.V., da Silva, F.T., de Paiva, T.C.B., Silva, G.C.: Efficient dairy wastewater treatment and power production using graphite cylinders electrodes as a biofilter in microbial fuel cell. Biochem. Eng. J. Eng. J. **178**, 108283 (2022)
4. Choudhury, P., Bhunia, B., Bandyopadhyay, T.K., Ray, R.N.: The overall performance improvement of microbial fuel cells connected in series with dairy wastewater treatment. J. Electrochem. Sci. Technol.Electrochem. Sci. Technol. **12**(1), 101–111 (2021)
5. Choudhury, P., Ray, R.N., Tiwari, O.N., Bandyopadhyay, T.K., Muthuraj, M., Bhunia, B.: Strategies for improvement of microbial fuel cell performance via stable power generation from real dairy wastewater. Fuel **288**, 119653 (2021)
6. Sar, T., Harirchi, S., Ramezani, M., Bulkan, G., Akbas, M.Y., Pandey, A., Taherzadeh, M.J.: Potential utilization of dairy industries by-products and wastes through microbial processes: a critical review. Sci. Total. Environ. **810**, 152253 (2022)
7. Roufou, S., Griffin, S., Katsini, L., Polańska, M., Van Impe, J.F., Valdramidis, V.P.: The (potential) impact of seasonality and climate change on the physicochemical and microbial properties of dairy waste and its management. Trends Food Sci. Technol. **116**, 1–10 (2021)
8. Awasthi, M.K., Paul, A., Kumar, V., Sar, T., Kumar, D., Sarsaiya, S., Taherzadeh, M.J.: Recent trends and developments on integrated biochemical conversion process for valorization of dairy waste to value added bioproducts: a review. Biores. Technol.. Technol. **344**, 126193 (2022)
9. Usmani, Z., Sharma, M., Gaffey, J., Sharma, M., Dewhurst, R.J., Moreau, B., Gupta, V.K.: Valorization of dairy waste and by-products through microbial bioprocesses. Bioresour. Technol. 126444 (2021)

10. McAteer, P.G., Trego, A.C., Thorn, C., Mahony, T., Abram, F., O'Flaherty, V.: Reactor configuration influences microbial community structure during high-rate, low-temperature anaerobic treatment of dairy wastewater. Biores. Technol.. Technol. **307**, 123221 (2020)
11. Mozejko-Ciesielska, J., Marciniak, P., Moraczewski, K., Rytlewski, P., Czaplicki, S., Zadernowska, A.: Cheese whey mother liquor as dairy waste with potential value for polyhydroxyalkanoate production by extremophilic Paracoccus homiensis. Sustain. Mater. Technol. **33**, e00449 (2022)
12. Kothari, R., Pathak, V.V., Kumar, V., Singh, D.P.: Experimental study for growth potential of unicellular alga Chlorella pyrenoidosa on dairy waste water: an integrated approach for treatment and biofuel production. Biores. Technol. Technol. **116**, 466–470 (2012)
13. Ferreira, T.F., Santos, P.A., Paula, A.V., de Castro, H.F., Andrade, G.S.: Biogas generation by hybrid treatment of dairy wastewater with lipolytic whole cell preparations and anaerobic sludge. Biochem. Eng. J. Eng. J. **169**, 107965 (2021)
14. Shen, J., Wang, C., Liu, Y., Hu, C., Xin, Y., Ding, N., Su, S.: Effect of ultrasonic pretreatment of the dairy manure on the electricity generation of microbial fuel cell. Biochem. Eng. J. Eng. J. **129**, 44–49 (2018)
15. Raja, S.W., Thanuja, K.G., Karthikeyan, S., Marimuthu, S.: Exploring the concurrent use of microalgae Coelastrella sp. for electricity generation and dairy wastewater treatment. Bioresour. Technol. Rep. **17**, 100889 (2022)
16. Goli, A., Shamiri, A., Khosroyar, S., Talaiekhozani, A., Sanaye, R., Azizi, K., Talaiekhozani, A.: A review on different aerobic and anaerobic treatment methods in dairy industry wastewater. A review on different aerobic and anaerobic treatment methods in dairy industry wastewater. J. Environ. Treat. Tech. **7**(1), 113–41 (2019)
17. Yonar, T., Sivrioğlu, Ö., Özengin, N.: Physico-chemical treatment of dairy industry wastewaters: a review. In: Koca, N. (ed.) Technological Approaches for Novel Applications in Dairy Processing, p. 179 (2018)
18. Johannesson, G.H., Crolla, A., Lauzon, J.D., Gilroyed, B.H.: Estimation of biogas co-production potential from liquid dairy manure, dissolved air flotation waste (DAF) and dry poultry manure using biochemical methane potential (BMP) assay. Biocatal. Agric. Biotechnol. Agric. Biotechnol. **25**, 101605 (2020)
19. Chu, C.Y., Wang, Z.F.: Dairy cow solid waste hydrolysis and hydrogen/methane productions by anaerobic digestion technology. Int. J. Hydrogen Energy **42**(52), 30591–30598 (2017)
20. Zhong, J., Stevens, D.K., Hansen, C.L.: Optimization of anaerobic hydrogen and methane production from dairy processing waste using a two-stage digestion in induced bed reactors (IBR). Int. J. Hydrogen Energy **40**(45), 15470–15476 (2015)
21. Lhanafi, S., Anfar, Z., Chebli, B., Benafqir, M., El Haouti, R., Azougarh, Y., El Alem, N.: Factorial experimental design to enhance methane production of dairy wastes co-digestion. Sustain. Environ. Res. **28**(6), 389–395 (2018)
22. Choudhury, P., Ray, R.N., Bandyopadhyay, T.K., Tiwari, O.N., Bhunia, B.: Kinetics and performance evaluation of microbial fuel cell supplied with dairy wastewater with simultaneous power generation. Int. J. Hydrogen Energy **46**(31), 16815–16822 (2021)
23. Selvasembian, R., Mal, J., Rani, R., Sinha, R., Agrahari, R., Joshua, I., Pradhan, N.: Recent progress in microbial fuel cells for industrial effluent treatment and energy generation: Fundamentals to scale-up application and challenges. Biores. Technol. Technol. **346**, 126462 (2022)
24. Bhargavi, G., Venu, V., Renganathan, S.: Microbial fuel cells: recent developments in design and materials. In: IOP Conference Series: Materials Science and Engineering, vol. 330, no. 1, p. 012034. (2018)
25. Sekar, A.D., Jayabalan, T., Muthukumar, H., Chandrasekaran, N.I., Mohamed, S.N., Matheswaran, M.: Enhancing power generation and treatment of dairy waste water in microbial fuel cell using Cu-doped iron oxide nanoparticles decorated anode. Energy **172**, 173–180 (2019)
26. Sanjay, S., Udayashankara, T.H.: Dairy wastewater treatment with bio-electricity generation using dual chambered membrane-less microbial fuel cell. Mater. Today: Proc. **35**, 308–311 (2021)

27. Tajdid, K.R., Aber, S., Nofouzi, K., Ebrahimi, S.: Treatment of mixed dairy and dye wastewater in anode of microbial fuel cell with simultaneous electricity generation. Environ. Sci. Pollut. Res. **27**(35), 43711–43723 (2020)
28. Mansoorian, H.J., Mahvi, A.H., Jafari, A.J., Khanjani, N.: Evaluation of dairy industry wastewater treatment and simultaneous bioelectricity generation in a catalyst-less and mediator-less membrane microbial fuel cell. J. Saudi Chem. Soc. **20**(1), 88–100 (2016)
29. Choudhury, P., Ray, R.N., Bandyopadhyay, T.K., Basak, B., Muthuraj, M., Bhunia, B.: Process engineering for stable power recovery from dairy wastewater using microbial fuel cell. Int. J. Hydrogen Energy **46**(4), 3171–3182 (2021)
30. Choudhury, P., Ray, R.N., Bandyopadhyay, T.K., Bhunia, B.: Fed batch approach for stable generation of power from dairy wastewater using microbial fuel cell and its kinetic study. Fuel **266**, 117073 (2020)
31. Choudhury, P., Bhunia, B., Mahata, N., Bandyopadhyay, T.K.: Optimization for the improvement of power in equal volume of single chamber microbial fuel cell using dairy wastewater. J. Indian Chem. Soc. **99**(6), 100489 (2022)
32. Vidhyeswari, D., Surendhar, A., Bhuvaneshwari, S.: Evaluation of power generation and treatment efficiency of dairy wastewater in microbial fuel cell using TiO_2–SPEEK as proton exchange membrane. Water Sci. Technol. **84**(10–11), 3388–3402 (2021)
33. Marassi, R.J., Queiroz, L.G., Silva, D.C., Dos Santos, F.S., Silva, G.C., de Paiva, T.C.: Long-term performance and acute toxicity assessment of scaled-up air–cathode microbial fuel cell fed by dairy wastewater. Bioprocess Biosyst. Eng. **43**(9), 1561–1571 (2020)
34. Santos, A.D., Martins, R.C., Quinta-Ferreira, R.M., Castro, L.M.: Moving bed biofilm reactor (MBBR) for dairy wastewater treatment. Energy Rep. **6**, 340–344 (2020)
35. Modi, J., Choumal, A., Vyas, D., Shah, D., Joshi, K., Patel, K., Iyer, K.: Sustainable technology for modern era effluent treatment: Microbial fuel cell. Mater. Today: Proc. **57**, 1781–1788 (2022)
36. Rani, G., Banu, J.R., Kumar, G., Yogalakshmi, K.N.: Statistical optimization of operating parameters of microbial electrolysis cell treating dairy industry wastewater using quadratic model to enhance energy generation. Int. J. Hydrogen Energy **47**(88), 37401–37414 (2022)
37. Elakkiya, E., Matheswaran, M.: Comparison of anodic metabolisms in bioelectricity production during treatment of dairy wastewater in Microbial Fuel Cell. Biores. Technol. **136**, 407–412 (2013)
38. Das, B., Sarkar, S., Sarkar, A., Bhattacharjee, S., Bhattacharjee, C.: Recovery of whey proteins and lactose from dairy waste: a step towards green waste management. Process. Saf. Environ. Prot.Saf. Environ. Prot. **101**, 27–33 (2016)
39. Martínez-Ruano, J.A., Restrepo-Serna, D.L., Carmona-Garcia, E., Giraldo, J.A.P., Aroca, G., Cardona, C.A.: Effect of co-digestion of milk-whey and potato stem on heat and power generation using biogas as an energy vector: techno-economic assessment. Appl. Energy **241**, 504–518 (2019)
40. Arias, A., Feijoo, G., Moreira, M.T.: Process and environmental simulation in the validation of the biotechnological production of nisin from waste. Biochem. Eng. J. 108105 (2021)
41. Ottaviano, L.M., Ramos, L.R., Botta, L.S., Amâncio Varesche, M.B., Silva, E.L.: Continuous thermophilic hydrogen production from cheese whey powder solution in an anaerobic fluidized bed reactor: Effect of hydraulic retention time and initial substrate concentration. Int. J. Hydrogen Energy **42**(8), 4848–4860 (2017)
42. Schierano, M.C., Panigatti, M.C., Maine, M.A., Griffa, C.A., Boglione, R.: Horizontal subsurface flow constructed wetland for tertiary treatment of dairy wastewater: removal efficiencies and plant uptake. J. Environ. Manage. **272**, 111094 (2020)
43. Yadav, J.S.S., Yan, S., Ajila, C.M., Bezawada, J., Tyagi, R.D., Surampalli, R.Y.: Food-grade single-cell protein production, characterization and ultrafiltration recovery of residual fermented whey proteins from whey. Food Bioprod. Process.Bioprod. Process. **99**, 156–165 (2016)

44. Iglesias-Iglesias, R., Portela-Grandío, A., Treu, L., Campanaro, S., Kennes, C., Veiga, M.C.: Co-digestion of cheese whey with sewage sludge for caproic acid production: role of microbiome and polyhydroxyalkanoates potential production. Biores. Technol. Technol. **337**, 125388 (2021)
45. Carvalho, F., Prazeres, A.R., Rivas, J.: Cheese whey wastewater: characterization and treatment. Sci. Total. Environ. **445–446**, 385–396 (2013)
46. Zotta, T., Solieri, L., Iacumin, L., Picozzi, C., Gullo, M.: Valorization of cheese whey using microbial fermentations. Appl. Microbiol. Biotechnol.Microbiol. Biotechnol. **104**, 2749–2764 (2020)
47. Mohamed, A.Y.A., Siggins, A., Healy, M.G., Uallacháin, D.Ó., Fenton, O., Tuohy, P.: A novel hybrid coagulation-constructed wetland system for the treatment of dairy wastewater. Sci. Total. Environ. **847**, 157567 (2020)

Application of Nanomaterials in Water Purification: A Thematic Review

Bishnu Kant Shukla, Rahul Dev Singh, Nitya Singh, Shivam Singh, Yash Kumar Burman, and Rahul Singh

Abstract Traditional methods of purifying water such as coagulation/flocculation and chlorine are being replaced by innovative concepts and technologies of nanofibers as they are not able to give satisfactory result in presence of organic matter in wastewater. Nanomaterials can be used in treatment applications of wastewater because of their large specific surface area, significant reactivity, high level of functionalization, size-reliant features, affinity for specific contaminants, etc. Nanomaterials can be used to produce membranes and filters that are energy and cost-efficient due to their good flux rates, selective permeability, reusability, and enhanced durability. Several different categories of nanomaterials including carbon nanotubes (CNTs), antimicrobial nanomaterials, nanosorbents, single enzyme nanoparticles (SENs), self-assembled monolayers on mesoporous silica (SAMMS), and dendrimers have been discussed in this paper along with their production processes and applications.

Keywords Nanofibers · Self-assembled monolayers on mesoporous silica (SAMMS) · Carbon nanotubes (CNTs) · Single enzyme nanoparticles (SENs) · Desalination

1 Introduction

The health of humans depends on access to clean water, which is defined as water devoid of germs and hazardous substances. Due to (i) prolonged droughts, (ii) population enlargement, (iii) stricter health-based laws, and (iv) conflicting user expectations, the globe is currently confronting severe problems in satisfying growing needs for clean water [1–3]. The nanomaterials applications in the water treatment which are polluted by bacteria, viruses, radionuclides, inorganic and organic solutes, toxic metal ions, and poisonous metal ions are briefly discussed in this article. One of the technologies that is developing the fastest is nanotechnology. Numerous prospects are

B. K. Shukla (✉) · R. D. Singh · N. Singh · S. Singh · Y. K. Burman · R. Singh
Department of Civil Engineering, JSS Academy of Technical Education, Noida 201301, India
e-mail: bishnukantshukla@gmail.com

S. Saxena et al. (eds.), *Emerging Materials and Technologies in Water Remediation and Sensing*, Lecture Notes in Civil Engineering 439,
https://doi.org/10.1007/978-981-99-6762-9_6

opening up for the advancement of purification systems for wastewater and water that are more economical and ecologically friendly thanks to developments in nanoscale science and engineering [4].

New procedures and systems as well as altered design production sequences are presented by science and technology of membrane. Compared to conventional separation techniques like filtration and sedimentation, membrane procedures are more effective. Reverse osmosis (RO), forward osmosis (FO), pervaporation (PV), membrane distillation (MD), ultrafiltration (UF), microfiltration (MF), and nanofiltration are the furthermost widely utilized technologies of membrane for treating brackish and salty water (NF) [5].

1.1 Water Transportation and Distribution

Although water covers more than 71% of surface of globe, there is still a relatively limited amount of freshwater that may be used directly. Since the oceans contain almost all the water on Earth, direct human usage is not possible. The usable freshwater makes up the remaining 3%. About 68.7% of this 3% is trapped in ice caps and glaciers. Nearly 30% of the rest is made up of groundwater, which is heavily contaminated with both organic pollutants as a result of home, agricultural, and disposal of industrial wastes. Only roughly 0.007% of the total water in this world is supplied by rivers and lakes for human use [4, 6].

1.2 Nanoparticles and Water Purification

The organic and inorganic wastes in water can be controlled, and a high level of immaculateness may be attained via the application of current water purification techniques. As harmful pollutants are avoided, membrane and filter procedures are superior to chemical ones; nonetheless, their main downside is membrane and filter fouling. Higher energy requirements for the pumping system are a result of membrane fouling. Chemical processes result in dangerous reactive compounds that are subsequently crushed or burned and dumped in landfills. These two techniques disrupt the natural system and are damaging to the ecosystem [1, 4].

Recent research suggests that by employing nanofiltration, nanoparticles, or other products originating from the nanotechnology development, many problems relating to water quality may be fixed or significantly improved. Figure 1 illustrates a three-circle model. Circle 3 showcases the applied innovations derived from research, like nanoparticles and nanofiltration methods, and their societal impact, particularly in the realm of water quality improvement. Circle 1 represents the fundamental building blocks, such as materials at the nanoscale, while Circle 2 denotes the translation of these materials into observed applications, underpinning the solutions for community-wide challenges. Among the developments of most promising and intriguing are those in creating of ground-breaking methodologies for desalination of water. Numerous institutes are investigating the use of certain nanoparticles, either on some structural media or embedded in membranes, to quickly, cheaply, and efficiently make unusable water drinkable. Large volumes of wastewater are produced by several companies. Industry would experience considerable cost, time, and labor savings by removing impurities and reusing the decontaminated water, which would enhance environmental stewardship. The ultimate objective is to be able to promptly, efficiently, and affordably remove harmful substances from subsurface and other settings that are very tough to admittance in situ [1, 8, 9].

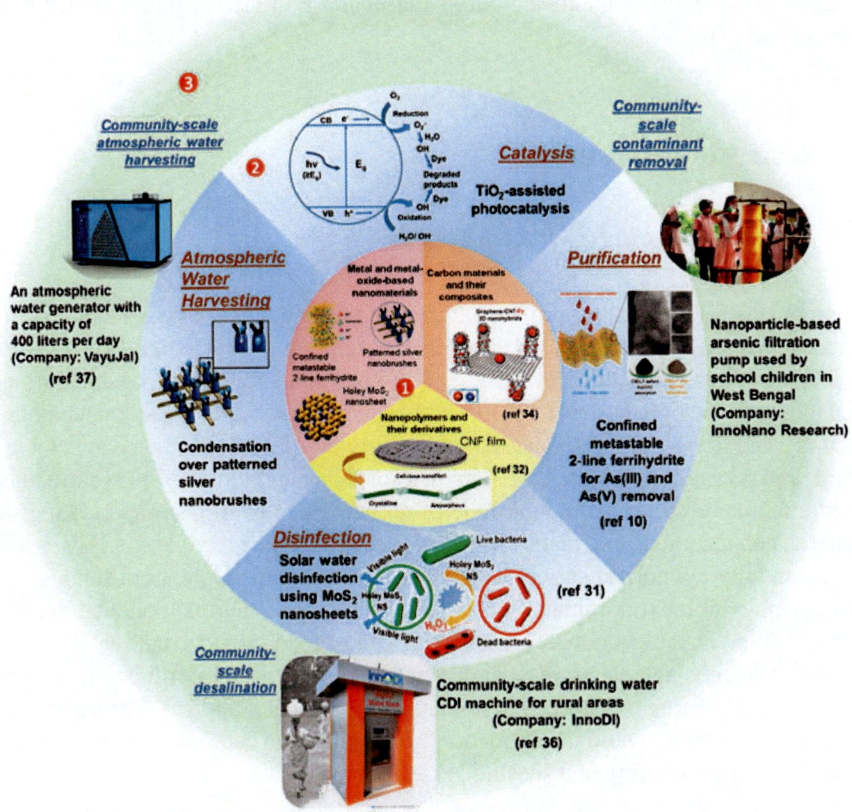

Fig. 1 The three-circle model of nanotechnology application and its societal impact [11, 12]

1.3 Nanophotocatalysis

The objective is to be able to quickly, effectively, and affordably remove harmful substances from subterranean and other areas that are very challenging to access in situ. Metal oxides like TiO_2, ZnO, SnO_2, and sulfides like ZnS, as well as nanoparticles, are effective oxidizable agents because they readily engross light near ultraviolet or in the visible regions of the electromagnetic spectrum, as are good photocatalysts.

Therefore, photocatalytic oxidation is a sophisticated oxidation method for eliminating microbiological pathogens and trace pollutants from wastewater. In order to improve the biodegradability of dangerous and non-biodegradable organic pollutants, it is also a helpful pre-treatment. The deferred kinetics brought on by the small fluency of light and activity of photocatalytic is the main hindrance to its implementation widespread [7].

2 Review of Literature

2.1 Nanotechnology Through Ages

Components for water filtration and purification have generally been manufactured of carbon throughout history and into the present. Activated (stimulated) carbon was first synthesized in the decade of 1940 and has undergone numerous alterations, including the use of nanostructures like carbon nanofibers (CNFs), carbon nanotubes (CNTs), and graphene-built materials [10]. With their impressive mechanical, electrical, and thermal attributes, CNFs, known for their unique high aspect ratio, have been effectively utilized in a multitude of domains like energy storage, composite material enhancement, and in the electronics sector. On the other hand, CNTs, which are essentially cylindrical structures formed by rolling up single-atom-thick sheets of carbon, exhibit extraordinary durability, electrical and thermal conduction capabilities. Their versatile nature has led to their widespread use in diverse areas such as materials science and electronics, and their potential in water purification contexts is increasingly being explored. These materials have been applied in treatment via membrane-based separation, disinfection, adsorption, and catalytic wet air oxidation as well as for monitoring and sensing [11].

2.2 Desalination

Desalination is a critical resolution to the shortage of water since over forty percent of the global population lives within one hundred kilometers marine. There are now 19,744 operational plants of desalination in 150 operational nations, providing 100

Mm^3 of water on daily basis to 300 million people globally. The process of desalination is still an energy-concentrated and environmentally risky process. There are three main obstacles for desalination technologies [11, 12]:

(1) high consumption of specific energy (SEC),
(2) fossil fuel combustion produces CO_2 emissions, and
(3) owing to the release of concentrated saltwater back into the water, adverse effects on marine ecosystems.

2.3 *Membrane-Based Desalination*

For treatment of water with TDS, generally, less than 45,000 mg/L, membrane-based separation is sufficient. Due to pointedly lower requirements of specific energy, such techniques (for example, forward osmosis, reverse osmosis, nanofiltration electrodialysis) have had marvelous viable accomplishments.

The greatest energy-consuming component of the process of desalination is the membrane module that accounts for about 71% (2.50–4.00 kWh/m^3 for salt-water RO) of the SEC [12–14]. The most potential low-energy alternatives include aligned carbon nanotube membranes, membranes of protein-polymer biomimetic, and membranes having thin-film nanocomposite, all of which may be developed using nanotechnology. The goal of many studies on membrane nanotechnologies is to combine nanomaterials with polymeric or inorganic membranes to produce synergistic effects; these materials are typically oxides of metal (TiO_2, zeolites, and Al_2O_3), some of which have activities of antimicrobial, photocatalytic, etc. (carbon, Ag nanotubes, etc.) [7]. Due to their suitability and affordability for assembly-line production, nanocomposites having polymeric membranes have achieved success commercially. Due to higher permeability and superior selectivity than more traditional membranes which are cellulose acetate-based, polyamide membranes are commonly utilized [11]. CNT-aligned membranes are appropriate for ultrafiltration. Functionalized CNT tips with aliphatic groups like Zwitter ionic species or carboxylic acids to upsurge their selectivity [15]. Existing membranes have become more durable, more permeable, and better at rejecting solutes because of the incorporation of CNTs [16]. A unique type of membrane that can dynamically cleanse water is called a nanocomposite membrane. It consists of a thin polymeric film surface that is painted or contains nanofillers. Nanocomposite membranes may be built using nanomaterials in a variety of shapes and dimensions. For example, nanoparticles are frequently utilized as nanofillers for polymeric membrane hydrophilization or mechanical reinforcement [17]. Detailed classifications of water desalination process have been presented in Fig. 2.

Tables 1 and 2 comparing the performance of nanotechnology-based and traditional reverse osmosis membranes, including mixed matrix (carbon nanotubes biomimetic), organic (aquaporin biomimetic), and inorganic (zeolite thin-film nanocomposite) [7], are also incorporated in this review.

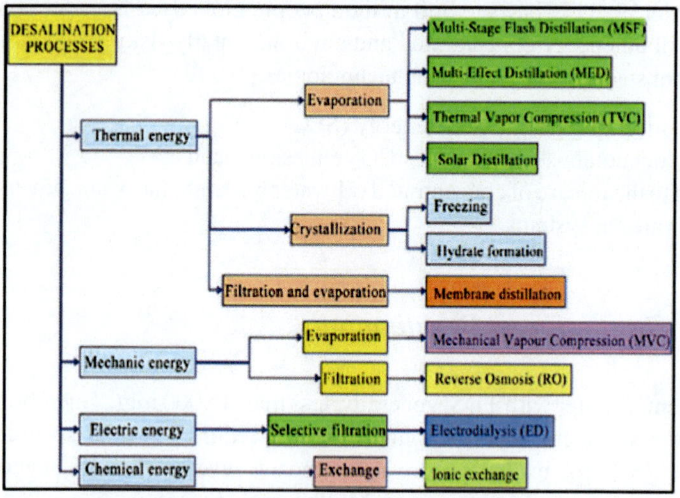

Fig. 2 Classification of water desalination processes [13]

2.4 *Reverse Osmosis*

Technique of desalinating osmosis producing clean water is called reverse osmosis (RO). In the process of reverse osmosis, a cooperative property, osmotic pressure, caused by chemical differences of potential of the solvent, a thermodynamic representative, is overwhelmed by an exerted pressure by application of a pressure superior to the osmotic pressure. By transporting a solvent across a semi-permeable membrane from an area of high concentration of solutes to an area of low solute concentration, reverse osmosis can be accomplished [18]. The four processes of a RO seawater desalination plant (as explained in Figs. 3 and 4) are high-pressure pumping, membrane separation, pre-treatment, and post-treatment [5]. Reverse osmosis is most often and effectively used to separate clean water from saltwater and brackish water. Traditional RO membranes for desalination are made of polymeric materials including polyamide spiral wound, asymmetric cellulose triacetate hollow fiber, aromatic polyamide-hydrazide, polypiperzine-amide, and polyoxadiazole, among others. Since the first plant of RO desalination, these polymeric RO membranes dominated commercial uses. They provide affordable fabrication due to their advanced technological capabilities [18].

2.5 *Electrodialysis*

An electric field is used in the electro-membrane desalination process known as electrodialysis (ED) to extract ions from water. Feed water flows into the channels between the exchange membranes for anion (AEMs) and exchange membranes

Table 1 Typical organic membranes, including permeation fluxes, technology, process types, and driving forces

Process	Driving force	Potential use	Typical flux	Membrane material	Temperature	References
RO	Induced pressure	Desalination	90 L/m^2 h	PA, CA	60 °C	[26–28]
FO	Osmosis	Desalination	3.1 L/m^2 h	Cellulose triacetate (CTA)	50 °C	[29–31]
MD	Thermal	Desalination, brine concentration	n.r	Polyvinylidene difluoride (PVDF), PTFE, and PP	50–80 °C	[10, 32–34]
PV	Thermal	Desalination	0.03 L/m^2 h	PVDF, Polyether amide	90 °C	[35, 36]
MF	Induced pressure	Desalination	n.r	PTFE, Acrylonitrile polymer	25 °C	[37, 38]
UF	Induced pressure	Desalination	91.2–103 L/m^2 h	PTFE, PVDF and CTA, regenerated cellulose	75 °C	[37, 39, 40]
NF	Induced pressure	Desalination	17.4 L/m^2 h	CTA	25–55 °C	[40, 41]
ED	Electric current	Brackish water desalination	3.5 L/m^2 h	CTA	Ambient temperature	[26, 27, 42]
MED	Thermal	Desalination, brine concentration	n.r	PVDF, PTFE, and PP	50 °C	[33]

for cation (CEMs), which are interchangeably located amid two electrodes. Ions are propelled toward electrodes with contradictory signs by the exerted electric field. Anions move via AEM but are hampered by CEM, whereas cations may flow through CEM but are prevented by AEM. A group of repeating unit cells with an AEM, a CEM, a concentrate channel, and a dilute channel make up a stack of ED [20]. IEMs used in ED need to be highly permselective, highly conductive, low in resistance, and chemically, mechanically, and dimensionally stable. Compounds having polymeric make-up with functional fixed-charge groups and mobile counterions make up EMs [21, 22]. The polymeric structure of IEMs has been made up of a variety of polymers, such as polypropylene (PP), polyethylene (PE), polyether ketone (PEK), PES, polyimide (PI), polybenzimidazole (PBI), polysulfide (PSU), poly(p-phenalene oxide), PEI and polyvinylidene fluoride (PVDF) [23, 24].

Table 2 Antimicrobial properties of nanomaterials

Nanomaterials	Microorganisms	References
Pure metals		
Au	**Bacteria**: Pseudomonas aeruginosa, Staphylococcus aureus, Salmonella typhi, Escherichia coli **Fungi**: Chrysosporium indicum, Candida albicans, Trichopyton rubrum	[50, 51]
Ag	**Bacteria**: Klebsiella pneumoniae, Escherichia coli, Pseudomonas aeruginosa, Salmonella typhi, Staphylococcus aureus, Proteus vulgaris **Fungi**: Trichopyton rubrum, Chrysosporium indicum, Candida albicans	[50, 52–54]
Bi	**Bacteria**: Streptococcus mutans	[55]
Cu	**Bacteria**: Staphylococcus epidermidis, Staphylococcus aureus **Fungi**: Candida albicans, Candida parapsilosis	[56]
Metal oxides		
CuO	**Bacteria**: Pseudomonas aeruginosa, Escherichia coli, Enterococcus faecalis, Klebsiella pneumonia, Salmonella typhimurium, Shigella flexneri, Staphylococcus aureus, Proteus vulgaris	[57]
TiO_2	**Bacteria**: Staphylococcus aureus, Escherichia coli, Bacillus subtilis, Pseudomonas aeruginosa **Fungi**: Candida albicans	[58]
Fe_3O_4	**Bacteria**: Proteus mirabilis, Escherichia coli, Bacillus subtilis	[59]
ZnO	**Bacteria**: Bacillus subtilis, Staphylococcus. aureus, Pseudomonas aeruginosa, Escherichia coli, Proteus mirabilis **Fungi**: Candida tropicalis, Candida albicans	[60]
Hybrids and composites		
Ag-Fe_3O_4 c-Fe_2O_3-Ag	**Bacteria**: Enterococcus faecalis, Staphylococcus aureus, Pseudomonas aeruginosa, Escherichia coli, Enterococcus faecium, Staphylococcus epidermidis, Klebsiella pneumoniae **Fungi**: Candida tropicalis, Candida albicans (I, II), Candida parapsilosis	[61]
Ag-Au	**Bacteria**: Escherichia coli, Staphylococcus aureus, Pseudomonas aeruginosa, Salmonella typhi **Fungi**: Candida albicans, Trichopyton rubrum, Chrysosporium indicum	[50]
Carbon-based nanomaterials		
SWCNTs	**Bacteria**: Escherichia coli	[62]
Fullerenes	**Bacteria**: Escherichia coli, Staphylococcus aureus, Pseudomonas aeruginosa **Fungi**: Candida albicans	[63]
Graphene oxides	**Bacteria**: Pseudomonas aeruginosa	[64]

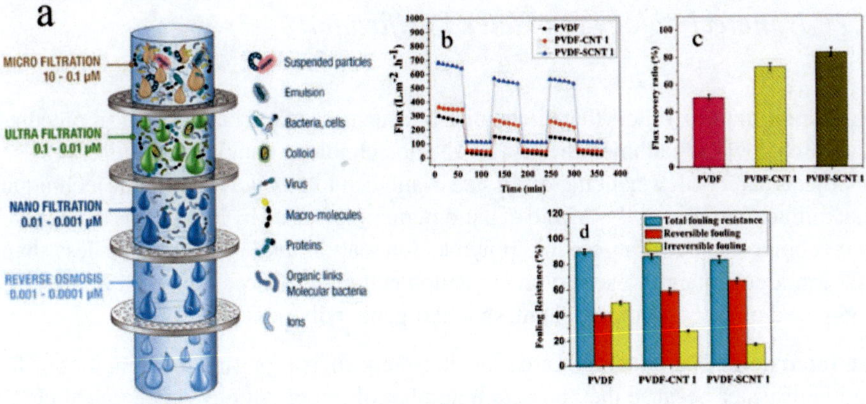

Fig. 3 Reverse osmosis, nanofiltration, ultrafiltration, microfiltration, and particle filtration are all depicted schematically [17]

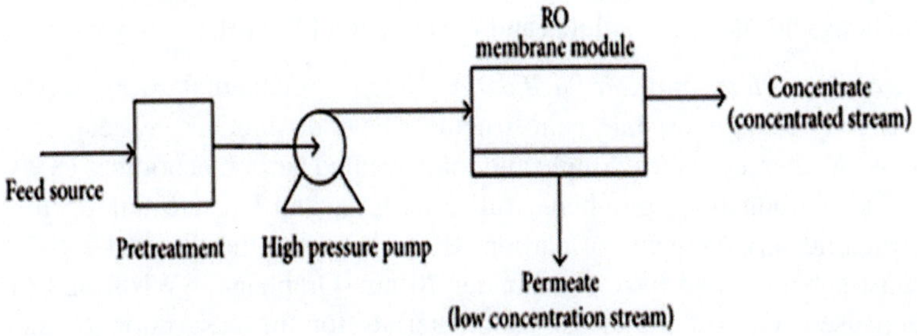

Fig. 4 Schematic diagram of RO process [19]

2.6 *Microfiltration/Ultrafiltration/Nanofiltration*

The size of the membrane hole essentially distinguishes these procedures. MF membranes, which have wider holes than UF membranes, which have significantly smaller pores, tend to only reject comparatively larger microorganisms. Both bacteria and soluble supermolecules can be rejected by UF. NF is also referred to as "loose RO" at times [25]. The development of MF and UF low-pressure membrane technologies for the purification and treatment of water began in the 1960s. Colloidal suspensions may be effectively separated and treated with ME, whereas moderately sized molecule solutions or suspensions are treated using UF [5]. NF, which has pores in the range of 0.50–5.00 nm, may disperse colloidal particles between RO and UF (Fig. 3). It is pressure-driven and produces high permeate flow while being able to sustain variance pressures in the range of 500–2,000 kPa, which are below operating RO pressures [25].

2.7 Nanotechnology for Water Purification

Nanomembranes. Filters that distinguish between gases and liquids at the molecular level are known as nanomembranes [22]. Nanotechnology entails the creation and use of objects barely a few nanometers in size. Nanotechnology is a biomimetic technique that mimics nature exactly. A kind of nanomembrane is used by living cells to operate. Nanocomposites with an organic polymer foundation and a thickness of less than 100 nm are frequently used to create nanomembranes. These nanomembranes are composed of silica nanoparticle mesh and organic polymers [18].

Nanoparticles. Nanomaterials are especially appealing as parting medium for purification of water because they possess a number of important physicochemical characteristics (Fig. 5). These have substantially higher areas of surface as compared to bulk particles on the basis of mass. Adding different chemical groups to nanomaterials can boost their affinity for a particular molecule. In aquatic solutions, they can also perform as high selectivity, capacity, and recyclable ligands for organic, radionuclides and inorganic solutes and negative ion [43, 44].

Organic Nanoparticles in Water Purification. Since the elemental carbon's hybridization state may change, organic nanostructures have distinctive physical, chemical, and electrical characteristics. Single and multi-walled carbon nanotubes (SWCNTs, MWCNTs), carbon dots, graphene, fullerene C_{540} and C_{60} are among the main nano-structural arrangements of carbon. CDs are carbon nanoparticles (NPs) that are almost spherical and have sizes under 10 nm. Graphene, SWNT, and MWNT are often used. One of the finest nanomaterials for the adsorption of inorganic and organic contaminants in polluted water, these materials exhibit outstanding adsorption properties.

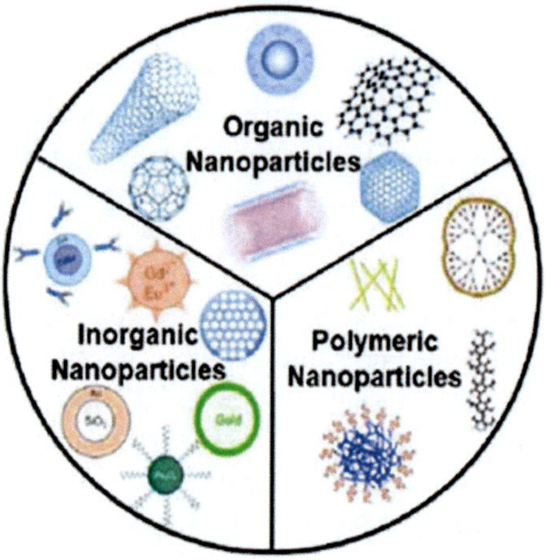

Fig. 5 Nanoparticle classification for water cleaning [44]

Inorganic Nanoparticles in Water Purification. Metal and their oxide nanoparticles (M&MO NPs) such as Ag NPs, TiO_2 NPs, CuO NPs, FeO and Fe NPs and nanoparticles of silica are the main inorganic nanoparticles employed in water treatment. M&MNO NPs exhibit outstanding heavy metal and chlorinated organic water pollution adsorption efficiency. Since both in and ex-situ applications must be adjusted, M&MO NPs cannot be avoided in cleanup. SiO_2 NPs are ideal for a wide variety of purposes due to the changeable pore diameter and simplicity of surface functionalization.

Polymeric Nanoparticles in Water Purification. An approach is to employ polymeric NPs. Here, a matrix or backing material is used to hold the NPs in place, preventing agglutination and enhancing the stability of pure nanoparticles. The size range for polymeric NPs is between 1 and 1000 nm. Numerous impurities including heavy metals like Hg, Fe As, and Mn; organic pollutants like medications, pesticides, organic volatile compounds, other aromatic and aliphatic compounds; gases like SO_2, NO_2, and CO_2; and microbes like viruses, bacteria, and other pathogens are all detected and removed using polymeric NPs.

2.8 Nanotechnology: Water Purification and Wastewater Treatment

Numerous potential applications for nanotechnology exist in the purification and treatment of wastewater. Membranes with better functionality, improved permeability to boost flow, and selective absorption of particular pollutants can be made using composite nanomaterials. Using "functional nanomaterials" to combine multiple processes, such as the degradation and separation of retentate, improves the efficacy of purification and waste treatment systems. Building high-performance point-of-use or small-scale systems with functional nanomaterials can help water delivery networks be more resilient [4]. The use of certain nanoparticles may quickly, cheaply, and effectively make potable water unfit, whether they are on membranes or other structured media. By eliminating impurities and reusing the purified water, industry would experience considerable cost, time, and labor savings, which would enhance environmental stewardship [1].

2.9 Reactive and Nanostructured Membranes

Reverse osmosis (RO), nanofiltration (NF), and ultrafiltration (UF) (Table 1) are developing into essential elements of cutting-edge desalination and water treatment technologies [40]. The effectiveness of using nanofiltration to desalinate water was examined [45, 46]. It was discovered that brackish water may be made drinkable by combining reverse osmosis with nanofiltration. Recently, Ref. [47] reported on the productive creation of carbon nanotube filters. These brand-new membranes for water filtration are made of hollow cylinders with walls made of radially oriented carbon nanotubes. He demonstrated how the filters were successful in getting rid of the poliovirus Sabin 1 and germs Staphylococcus aureus and Escherichia coli from polluted water. The effective production of alumina UF membranes utilizing alumina was reported by Ref. [48]. The "uniformity" of the alumina nanoparticles has a significant impact on the molecular weight cut-off (MWCO) and pore size of the membranes [1].

2.10 Antimicrobial Nanomaterials

Natural antibacterial chemicals, metals and their oxides and innovative manufactured nanomaterials are the three broad groups into which the antibacterial nanoparticles are divided [49]. There are several different ways in which interaction among nanoparticles and microbial cells takes place. The nanoparticles may directly intermingle with the microbial cells, such as by interfering with transmembrane electron transport, rupturing or piercing the envelope of cell or by oxidizing cell components, or they may generate byproducts that harm the cells [4]. Table 2 provides a thorough account of the antimicrobial activities displayed by several nanomaterial classes on fungi and bacteria.

2.11 Nanocomposite Membranes and Filters

Modern water filtration, sewage treatment, and desalination technologies are thought to rely most heavily on processes employing membrane and filter-based technologies. Nanoparticles, carbon nanotubes, zeolites, and dendrimers help to create water filtration systems that are less expensive and more effective [65]. Two categories of nanotechnology applications in membranes can be distinguished: nanoreactive filters and membranes, in which specially functionalized nanoparticles enhance the process of filtration by selectively interacting with the aimed impurities and nanostructured strainers, in which nanocapillary arrays or carbon nanotubes serve as the structural foundation for physical separation through nanofiltration. Examples of membranes made from nanomaterials include those made from carbon and inorganic ceramics, as well as those made by nanomaterial sintering, membrane reactors, and polymer nanocomposite membranes incorporating functional nanoparticles [65].

CNT Boosted Membranes. Due to their extremely tiny sizes, carbon nanotubes are excellent for physical separation operations. Nanoscale holes can be found in filters and membranes made with CNTs as reinforcement materials for filler. Water molecules may easily and quickly travel through the CNTs, whereas nearly all other pollutants are blocked [66]. Different CNT membranes include

(1) CNT—Silicon Nitride Membranes.
(2) CNT/Polymer Membranes.
(3) CNT Polymer Network Membranes.

Nanofilters. Water and salt may be reused through filtration procedures, which also reduce the quantity of effluent that must be emitted. Nanofilters provide effective water treatment by carrying out many tasks, high rates of salt and dye rejection, and low impelling pressure requirements. TiO_2-SiO_2 nanoparticles were used as grouting material in matrix of polyethylene glycol (PEG) to create a nanofilter. While SiO_2 nanoparticles operate as filler, resulting in the filter with nanoscale holes, TiO_2 nanoparticles are photocatalytic and assist in the destruction of organic contaminants and microorganisms [1, 67, 68].

Other Methods. Numerous nanomaterials are employed including Single Enzyme Nanoparticles (SENs), Self-Assembled Monolayers on Mesoporous Silica (SAMMS), and nanosorbents (MWNTs and Zeolites).

SAMMS. SAMMS are hexagonal honeycomb-shaped functionalized nanoporous ceramics. Because SAMMS may change the uncovered monolayer functional group, this variety of extremely sorbent ingredients has the potential to muddle a wide molecule range, including toxins and pollutants, particularly in the remediation of groundwater [63]. The three-stage process for making SAMMS includes (i) creating micelle templates having ordered crystalline liquid structure; (ii) creation of mesoporous anchors by precipitating oxides on the surface of micelle; (iii) removal of the surfactants by calcining of the organic-oxide material; and (iv) formation of monolayers within the mesopores by adding functionalized bifunctional silanes [4].

SENs. It is possible to create Single Enzyme Nanoparticles (SENs) that provide a more effective means of enzyme stabilizing. SENs are armored enzymes encased in an enclosure of defense that is only a few nanometers thick. Three steps are involved in the creation of SEN: (i) modification of the enzyme surface covalent through the dissolution in solvent (non-polar) and adding a vinyl group; (ii) production of polymers (vinyl group) with free trimethoxysilane groups associated with the enzyme surface; and (iii) formation of a few nanometers' thick silicate shell by hydrolysis of trimethoxysilane groups [1, 64–67].

Nanosorbents. In order to remove both inorganic and organic contaminants from polluted water, sorbents are frequently utilized as separating media in the process. Two crucial characteristics of nanoparticles make them particularly desirable as sorbents. These have considerably bigger areas of surface than bulk particles. The sorption of Cu(II), Pb(II), and Cd(II) onto multi-walled carbon nanotubes (MWCNTs) [69]. At room temperature, they found maximal sorption capacities for Cu(II) at 24.49 mg/g, Pb(II) at 97.08 mg/g and Cd(II) at 10.86 mg/g. Zeolites work well as ion-exchange media for metal ions and as sorbents. The density of Na ion-exchange sites is quite high in NaP1 zeolites (Na_6Al_6 $Si_{10}O_{32}$, $12H_2O$). By hydrothermally activating fly ash with a low Si/Al ratio at 150 °C in 1.0–2.0 M NaOH solutions, they may be produced at a reasonable cost [1].

3 Conclusions

Human health depends on clean water, which is also a crucial raw material in many important sectors, including the food, drug, and electronics industries. The inorganic and organic wastes originated from water can be controlled by current wastewater treatment and water purification techniques. But because of non-reusable filters and membranes, the difficulty of entirely purifying water, the impossibility to reuse the retentate, etc., these technologies are uneconomical and energy-exhaustive. The best technologies are those that can benefit from brackish waters' comparatively low salinity and produce water at high rates of recovery, which reduces production of concentrate. Membrane desalination technologies have been thoroughly investigated for brackish water treatment due to their typically low-energy needs. Nanotechnology and other cutting-edge techniques are quickly displacing older ideas. Nanomaterials have several important physicochemical characteristics that make them mostly appealing as media of separation for purification of water because they focus on refining the prevailing methods by enhancing the effectiveness of the methods and increasing the nanomaterials reusability, which saves on the cost of operating the plant/process.

This study reviews several nanomaterials and their use in water filtration. As greater advancement is made in the synthesis of reasonably priced and ecologically adequate functional nanomaterials, it is predicted that nanomaterials will play a crucial role in industrial and public water filtration systems.

References

1. Savage, N., Diallo, M.S.: Nanomaterials and water purification: opportunities and challenges. J. Nanopart. Res. **7**(4), 331–342 (2005)
2. Shukla, B.K., Gupta, A., Sharma, P.K., Bhowmik, A.R.: Pollution status and water quality assessment in pre-monsoon season: a case study of rural villages in Allahabad district, Uttar Pradesh, India. Mater. Today: Proc. **32**, 824–830 (2020)
3. Shukla, B.K., Sharma, P.K., Khan, M.A.: Physico-chemical study of some surface water bodies of Punjab. In: Journal of Physics: Conference Series, vol. 1531, no. 1, p. 012121. IOP Publishing (2020)
4. Dhakras, P.A.: Nanotechnology applications in water purification and wastewater treatment: a review. In: International Conference on Nanoscience, Engineering, and Technology (ICONSET 2011), pp. 285–291. IEEE (2011)
5. Aliyu, U.M., Rathilal, S., Isa, Y.M.: Membrane desalination technologies in water treatment: a review. Water Pract. Technol. **13**(4), 738–752 (2018)
6. Shukla, B.K., Khan, A., Saikiran, G., Sriram, M.A.: Comparative study on effect of variation in opening shape on oxygenation performance of surface jet aerators used in water and wastewater treatment. J. Green Eng. **9**(3), 427–440 (2019)
7. Villaseñor, M., Ríos, Á.: Nanomaterials for water cleaning and desalination, energy production, disinfection, agriculture, and green chemistry. Environ. Chem. Lett. **16**(1), 11–34 (2018)
8. Shenoy, A., Shukla, B.K., Bansal, V.: Sustainable design of textile industry effluent treatment plant with constructed wetland. Mater. Today: Proc. **61**, 537–542 (2022)
9. Shenoy, A., Bansal, V., Shukla, B.K.: Treatability of effluent from small scale dye shop using water hyacinth. Mater. Today: Proc. **61**, 579–586 (2022)

10. Pollard, S.J.T., Fowler, G.D., Sollars, C.J., Perry, R.: Low-cost adsorbents for waste and wastewater treatment: a review. Sci. Total Environ. **116**(1–2), 31–52 (1992)
11. Nagar, A., Pradeep, T.: Clean water through nanotechnology: needs, gaps, and fulfillment. ACS Nano **14**(6), 6420–6435 (2020)
12. Al-Ansari, A., Ettouney, H., El-Dessouky, H.: Water–zeolite adsorption heat pump combined with single effect evaporation desalination process. Renew. Energy **24**(1), 91–111 (2001)
13. Kabeel, A.E., Hamed, M.H., Omara, Z.M., Sharshir, S.W.: Water desalination using a humidification-dehumidification technique—a detailed review. Desalin. Water Treat. **52**(25–27), 4622–4640 (2013)
14. Ghurye, G.L., Mishra, D., Lucas, L.: Thermal desalination of produced water—an analysis of the partitioning of constituents into product streams and its implications for beneficial use outside the O&G industry. Water **13**(8), 1068 (2021)
15. Vahdatifar, S., Khodadadi, A.A., Mortazavi, Y., Greenlee, L.F.: Functionalized open-ended vertically aligned carbon nanotube composite membranes with high salt rejection and enhanced slip flow for desalination. Sep. Purif. Technol. **279**, 119773 (2021)
16. Ali, S., Rehman, S.A.U., Luan, H.Y., Farid, M.U., Huang, H.: Challenges and opportunities in functional carbon nanotubes for membrane-based water treatment and desalination. Sci. Total Environ. **646**, 1126–1139 (2019)
17. Yu, S., Tang, H., Zhang, D., Wang, S., Qiu, M., Song, G., Fu, D., Hu, B., Wang, X.: MXenes as emerging nanomaterials in water purification and environmental remediation. Sci. Total Environ. **811**, 152280 (2021)
18. Rho, H., Yu, P., Zhao, Z., Lee, C.S., Chon, K., Perreault, F., Alvarez, P.J., Amy, G., Westerhoff, P.: Inhibition of biofouling on reverse osmosis membrane surfaces by germicidal ultraviolet light side-emitting optical fibers. Water Res. **224**, 119094 (2022)
19. Jiang, A., Ding, Q., Wang, J., Jiangzhou, S., Cheng, W., Xing, C.: Mathematical modeling and simulation of SWRO process based on the simultaneous method. J. Appl. Math. (2014)
20. Honarparvar, S., Zhang, X., Chen, T., Alborzi, A., Afroz, K., Reible, D.: Frontiers of membrane desalination processes for brackish water treatment: a review. Membranes **11**(4), 246 (2021)
21. Strathmann, H.: Electrodialysis, a mature technology with a multitude of new applications. Desalination **264**, 268–288 (2010)
22. Warsinger, D.M., Chakraborty, S., Tow, E.W., Plumlee, M.H., Bellona, C., Loutatidou, S., Karimi, L., Mikelonis, A.M., Achilli, A., Ghassemi, A., Padhye, L.P.: A review of polymeric membranes and processes for potable water reuse. Prog. Polym. Sci. **81**, 209–237 (2018)
23. Hagesteijn, K.F., Jiang, S., Ladewig, B.P.: A review of the synthesis and characterization of anion exchange membranes. J. Mater. Sci. **53**(16), 11131–11150 (2018)
24. Merle, G., Chairuna, A., Ven, E.V.D., Nijmeijer, K.: An easy method for the preparation of anion exchange membranes: Graft-polymerization of ionic liquids in porous supports. J. Appl. Polym. Sci. **129**(3), 1143–1150 (2013)
25. Chollom, M.N., Rathilal, S., Alfa, D., Pillay, V.L.: The applicability of nanofiltration for the treatment and reuse of textile reactive dye effluent. Water Sa **41**(3), 398–405 (2015)
26. Greenlee, L.F., Lawler, D.F., Freeman, B.D., Marrot, B., Moulin, P.: Reverse osmosis desalination: water sources, technology, and today's challenges. Water Res. **43**(9), 2317–2348 (2009)
27. Pérez-González, A., Urtiaga, A.M., Ibáñez, R., Ortiz, I.: State of the art and review on the treatment technologies of water reverse osmosis concentrates. Water Res. **46**(2), 267–283 (2012)
28. Antony, A., Branch, A., Leslie, G., Le-Clech, P.: Impact of membrane ageing on reverse osmosis performance–Implications on validation protocol. J. Membr. Sci. **520**, 37–44 (2016)
29. Wang, W., Zhang, Y., Esparra-Alvarado, M., Wang, X., Yang, H., Xie, Y.: Effects of pH and temperature on forward osmosis membrane flux using rainwater as the makeup for cooling water dilution. Desalination **351**, 70–76 (2014)
30. Sahebi, S., Phuntsho, S., Kim, J.E., Hong, S., Shon, H.K.: Pressure assisted fertiliser drawn osmosis process to enhance final dilution of the fertiliser draw solution beyond osmotic equilibrium. J. Membr. Sci. **481**, 63–72 (2015)

31. Majeed, T., Phuntsho, S., Jeong, S., Zhao, Y., Gao, B., Shon, H.K.: Understanding the risk of scaling and fouling in hollow fiber forward osmosis membrane application. Process Saf. Environ. Prot. **104**, 452–464 (2016)
32. Alkhudhiri, A., Darwish, N., Hilal, N.: Membrane distillation: a comprehensive review. Desalination **287**, 2–18 (2012)
33. Maab, H., Al Saadi, A., Francis, L., Livazovic, S., Ghafour, N., Amy, G.L., Nunes, S.P.: Polyazole hollow fiber membranes for direct contact membrane distillation. Ind. Eng. Chem. Res. **52**(31), 10425–10429 (2013)
34. Wang, P., Chung, T.S.: Recent advances in membrane distillation processes: membrane development, configuration design and application exploring. J. Membr. Sci. **474**, 39–56 (2015)
35. Swenson, P., Tanchuk, B., Gupta, A., An, W., Kuznicki, S.M.: Pervaporative desalination of water using natural zeolite membranes. Desalination **285**, 68–72 (2012)
36. Zhu, B., Hong, Z., Milne, N., Doherty, C.M., Zou, L., Lin, Y.S., Hill, A.J., Gu, X., Duke, M.: Desalination of seawater ion complexes by MFI-type zeolite membranes: temperature and long term stability. J. Membr. Sci. **453**, 126–135 (2014)
37. Kennedy, M.D., Kamanyi, J., Rodríguez, S.S., Lee, N.H., Schippers, J.C., Amy, G.: Water treatment by microfiltration and ultrafiltration. Adv. Membr. Technol. Appl., 131–170 (2008)
38. Bush, J.A., Vanneste, J., Cath, T.Y.: Membrane distillation for concentration of hypersaline brines from the Great Salt Lake: effects of scaling and fouling on performance, efficiency, and salt rejection. Sep. Purif. Technol. **170**, 78–91 (2016)
39. Busch, M., Chu, R., Rosenberg, S.: Novel trends in dual membrane systems for seawater desalination: minimum primary pretreatment and low environmental impact treatment schemes. IDA J. Desalination Water Reuse **2**(1), 56–71 (2010)
40. Chekli, L., Phuntsho, S., Kim, J.E., Kim, J., Choi, J.Y., Choi, J.S., Kim, S., Kim, J.H., Hong, S., Sohn, J., Shon, H.K.: A comprehensive review of hybrid forward osmosis systems: performance, applications and future prospects. J. Membr. Sci. **497**, 430–449 (2016)
41. Noghabi, M.S., Razavi, S.M.A., Mousavi, S.M., Elahi, M., Niazmand, R.: Effect of operating parameters on performance of nanofiltration of sugar beet press water. Procedia Food Sci. **1**, 160–164 (2011)
42. Burn, S., Hoang, M., Zarzo, D., Olewniak, F., Campos, E., Bolto, B., Barron, O.: Desalination techniques—a review of the opportunities for desalination in agriculture. Desalination **364**, 2–16 (2015)
43. Mohamed, M.B.: Low Cost Nanomaterials for Water Desalination and Purification. United Nations UNSCO (2011)
44. Ajith, M.P., Rajamani, P.: Nanotechnology for water purification-current trends and challenges. J. Nanotechnol. Nanomater. **2**(2), 88–91 (2021)
45. Sandia, D.: Water Purification Roadmap–A Report of the Executive Committee, US Department of the Interior, Bureau of Reclamation and Sandia National Laboratories. Report2003 (2003)
46. Mohsen, M.S., Jaber, J.O., Afonso, M.D.: Desalination of brackish water by nanofiltration and reverse osmosis. Desalination **157**(1–3), 167 (2003)
47. Srivastava, A., Srivastava, O.N., Talapatra, S., Vajtai, R., Ajayan, P.M.: Carbon nanotube filters. Nat. Mater. **3**(9), 610–614 (2004)
48. DeFriend, K.A., Wiesner, M.R., Barron, A.R.: Alumina and aluminate ultrafiltration membranes derived from alumina nanoparticles. J. Membr. Sci. **224**(1–2), 11–28 (2003)
49. Li, Q., Mahendra, S., Lyon, D.Y., Brunet, L., Liga, M.V., Li, D., Alvarez, P.J.: Antimicrobial nanomaterials for water disinfection and microbial control: potential applications and implications. Water Res. **42**(18), 4591–4602 (2008)
50. Yallappa, S., Manjanna, J., Dhananjaya, B.L.: Phytosynthesis of stable Au, Ag and Au–Ag alloy nanoparticles using J. sambac leaves extract, and their enhanced antimicrobial activity in presence of organic antimicrobials. Spectrochim. Acta Part A: Mol. Biomol. Spectrosc. **137**, 236–243 (2015)
51. Bindhu, M.R., Umadevi, M.: Silver and gold nanoparticles for sensor and antibacterial applications. Spectrochim. Acta Part A Mol. Biomol. Spectrosc. **128**, 37–45 (2014)

52. Ahmed, M.J., Murtaza, G., Mehmood, A., Bhatti, T.M.: Green synthesis of silver nanoparticles using leaves extract of Skimmia laureola: characterization and antibacterial activity. Mater. Lett. **153**, 10–13 (2015)
53. Ahluwalia, V., Kumar, J., Sisodia, R., Shakil, N.A., Walia, S.: Green synthesis of silver nanoparticles by Trichoderma harzianum and their bio-efficacy evaluation against Staphylococcus aureus and Klebsiella pneumonia. Ind. Crops Prod. **55**, 202–206 (2014)
54. Ravindran, A., Chandran, P., Khan, S.S.: Biofunctionalized silver nanoparticles: advances and prospects. Colloids Surf., B **105**, 342–352 (2013)
55. Hernández Delgadillo, R., Velasco Arias, D., Diaz, D., Arévalo Niño, K., Garza Enriquez, M., De la Garza Ramos, M.A., Cabral Romero, C.: Zerovalent bismuth nanoparticles inhibit Streptococcus mutans growth and formation of biofilm. Int. J. Nanomedicine, 2109 (2012)
56. Kruk, T., Szczepanowicz, K., Stefańska, J., Socha, R.P., Warszyński, P.: Synthesis and antimicrobial activity of monodisperse copper nanoparticles. Colloids Surf. B: Biointerfaces **128**, 17–22 (2015)
57. Ahamed, M., Alhadlaq, H.A., Khan, M.A., Karuppiah, P., Al-Dhabi, N.A.: Synthesis, characterization, and antimicrobial activity of copper oxide nanoparticles. J. Nanomater., 17–17 (2014)
58. Pişkin, S., Palantöken, A., Yılmaz, M.S.: Antimicrobial activity of synthesized TiO_2 nanoparticles. In: International Conference on Emerging Trends in Engineering and Technology (ICETET'2013), Thailand (2013)
59. Arokiyaraj, S., Saravanan, M., Prakash, N.U., Arasu, M.V., Vijayakumar, B., Vincent, S.: Enhanced antibacterial activity of iron oxide magnetic nanoparticles treated with Argemone mexicana L. leaf extract: an in vitro study. Mater. Res. Bull. **48**(9), 3323–3327 (2013)
60. Elumalai, K., Velmurugan, S.: Green synthesis, characterization and antimicrobial activities of zinc oxide nanoparticles from the leaf extract of Azadirachta indica (L.). Appl. Surf. Sci. **345**, 329–336 (2015)
61. Prucek, R., Tuček, J., Kilianová, M., Panáček, A., Kvítek, L., Filip, J., Kolář, M., Tománková, K., Zbořil, R.: The targeted antibacterial and antifungal properties of magnetic nanocomposite of iron oxide and silver nanoparticles. Biomaterials **32**(21), 4704–4713 (2011)
62. Vecitis, C.D., Zodrow, K.R., Kang, S., Elimelech, M.: Electronic-structure-dependent bacterial cytotoxicity of single-walled carbon nanotubes. ACS Nano **4**(9), 5471–5479 (2010)
63. Tegos, G.P., Demidova, T.N., Arcila-Lopez, D., Lee, H., Wharton, T., Gali, H., Hamblin, M.R.: Cationic fullerenes are effective and selective antimicrobial photosensitizers. Chem. Biol. **12**(10), 1127–1135 (2005)
64. Gurunathan, S., Han, J.W., Dayem, A.A., Eppakayala, V., Kim, J.H.: Oxidative stress-mediated antibacterial activity of graphene oxide and reduced graphene oxide in Pseudomonas aeruginosa. Int. J. Nanomed. **7**, 5901 (2012)
65. Dey, T.: Nanotechnology for Water Purification. Universal-Publishers (2012)
66. Noy, A., Park, H.G., Fornasiero, F., Holt, J.K., Grigoropoulos, C.P., Bakajin, O.: Nanofluidics in carbon nanotubes. Nano Today **2**(6), 22–29 (2007)
67. Watlington, K.: Emerging Nanotechnologies for Site Remediation and Wastewater Treatment. Environmental Protection Agency, Washington DC (2005)
68. Zhong, L.S., Hu, J.S., Liang, H.P., Cao, A.M., Song, W.G., Wan, L.J.: Self-Assembled 3D flowerlike iron oxide nanostructures and their application in water treatment. Adv. Mater. **18**(18), 2426–2431 (2006)
69. Li, Y.H., Ding, J., Luan, Z., Di, Z., Zhu, Y., Xu, C., Wu, D., Wei, B.: Competitive adsorption of Pb^{2+}, Cu^{2+} and Cd^{2+} ions from aqueous solutions by multiwalled carbon nanotubes. Carbon **41**(14), 2787–2792 (2003)

Experimental Study on Green Synthesis of Copper Nanoparticles and Its Utilization for Degradation of Organic Dye

Sarvesh Datar, Shreyas Kadu, Pratiksha Khurpade, and Somnath Nandi

Abstract Metal and metal oxide nanoparticles are widely used to remove organic dyes from wastewater due to their higher surface area, high adsorption capacity, chemical stability, and photocatalytic activity. In present work, *Azadirachta indica* (neem) leaf extract was used as a reducing and capping agent for the green synthesis of copper nanoparticles (CuNPs) with homogeneous precipitation process. The developed green method of *Azadirachta indica* leaves extract was found to be simple, environment-friendly, and cost-effective as compared to existing toxic reducing agents. Synthesized CuNPs using green route were characterized by UV–visible spectroscopy, Fourier Transform Infrared spectroscopy (FTIR), and Scanning Electron Microscopy (SEM) analysis. UV–visible spectroscopy analysis showed characteristics absorption peaks at 800 nm confirming the formation of copper nanoparticles. FTIR spectra exhibited the presence of bioactive molecules in leaf extract which are responsible for the reduction of copper ions to CuNPs. Moreover, Scanning Electron Microscopic (SEM) analysis also revealed that synthesized CuNPs were in spherical shape with average size of particles ranging between 50 and 70 nm. Further, this study reports utilization of biosynthesized CuNPs for degradation of organic dye, namely, methylene blue dye from wastewater samples under sunlight. Result analysis revealed that approximately 60% of degradation of Methylene blue dye was observed within 120 min using green synthesized copper nanoparticles that exhibit good photocatalytic activity. This study demonstrates that green synthesized copper nanoparticles can be effectively used for removal of hazardous dye from industrial wastewater which contributes to cost-effective, environmental, and sustainable processes.

S. Datar · S. Kadu · P. Khurpade (✉)
School of Chemical Engineering, Dr. Vishwanath Karad MIT World Peace University, Paud Road, Kothrud, Pune 411038, India
e-mail: pratiksha.khurpadekale@mitwpu.edu.in

S. Nandi
Department of Technology, Savitribai Phule Pune University, Ganeshkhind, Pune 411 007, India

S. Saxena et al. (eds.), *Emerging Materials and Technologies in Water Remediation and Sensing*, Lecture Notes in Civil Engineering 439,
https://doi.org/10.1007/978-981-99-6762-9_7

Keywords Copper nanoparticles · Neem leaf extract · Methylene blue degradation · Wastewater

1 Introduction

Water pollution has become a great concern worldwide due to accidental or arbitrarily dumping of polluted wastewater from industries into aquatic systems [1]. Organic dyes are extensively used in various industries such as paper, cosmetics, textile, plastics, and leather tanning. These synthetic dyes are highly toxic, carcinogenic, and non-biodegradable due to their complex chemical structure [2]. However, the release of such harmful dyes into the water bodies without appropriate treatment can cause environmental problems along with health problems in human beings such as vomiting, diarrhea, skin diseases, cancer, allergic reactions, and mutation for people [3]. According to "The United Nations World Water Development Report 2017," over 800,000 deaths worldwide were caused by contaminated drinking water [4]. Hence, it is imperative to remove methylene blue dye from industrial wastewater before discharge into aquatic systems.

Different physical, chemical, and biological methods for the removal of dye from wastewater have been reported in the literature but most of them have certain demerits, such as requiring high temperature, harmful chemicals, high operating costs, and time-consuming [5]. In contrast, in recent years, green synthesized metal nanoparticles (NPs) have been employed in wastewater treatment along with medicine, drug delivery, and optoelectronics applications [6]. Moreover, antifungal, antibacterial, photocatalytic, properties are the important features of various metal nanoparticles [7]. Photocatalysts have been gaining more attention for the treatment of organic pollutants from past several years [1, 8]. Photocatalysis is a green method for the degradation of hazardous dyes in wastewater via sunlight or visible light-induced reaction with catalyst (i.e. nanoparticles) that produce free radical such as hydroxyl radical (OH^-). These radicals can break down the organic dyes molecules and degrade them efficiently. It eliminates the need for hazardous chemicals and thus resulting in more sustainable, efficient, stable, and eco-friendly approach as compared to conventional treatment methods [3, 9].

In the last few years, different approaches for synthesis of metal NPs have been invented and reported, for instance sol gel, hydrothermal, chemical reduction, chemical vapor deposition, etc. [10]. These methods are energy intensive and employ the use of toxic, hazardous, and expensive chemicals which limits their use for wide implementation [11]. Nowadays, green method utilizing plant extract for NPs synthesis is more widely used as compared to conventional methods as these sources are inexpensive, easily and abundantly available, stable and nontoxic. In addition, active components of leaf extract namely polyphenols, flavonoids, alkaloids, and proteins play a vital role in the reduction of metal ions to nanoparticles and also act as capping agent to prevent agglomeration of nanoparticles [10].

Fig. 1 Dried *Azadirachta indica* (neem) leaves

In past few years, metal NPs synthesized via green route have been successfully applied in organic dye removal because of its less toxicity, easy separation, and its photocatalytic activity [5, 12, 13]. Among various metal NPs, copper nanoparticles (CuNPs) have been receiving considerable attention due to their attractive features including low cost, easy availability, catalytic properties, high electrical conductivity, optical and antibacterial properties [14]. In literature, copper-based nanoparticles have been synthesized from several plant and fruit extracts such as Jatropha curcas leaves, Celastrus paniculatus Willd. Leaves, Piper retrofractum Vahl fruit extract, Ficus carica extract [3, 6, 9, 14]. However, green synthesis of CuNPs using *Azadirachta indica* leaf (more commonly known as neem leaf) extract is reported in very few previous research studies [10, 11]. Figure 1 shows the image of dried neem leaf. It belongs to Meliaceae family and available abundantly in India and in nearby subcontinents. Products from neem have good antibacterial, antifungal, antiviral properties and also help in controlling the blood sugar level [12]. Therefore, the present work focused on green synthesis of CuNPs utilizing neem leaves extract as a reducing and capping agent.

The objective of this work is to synthesize copper nanoparticles (CuNPs) by using leaf extract of *Azadirachta indica* and then evaluate their photocatalytic activity for the decomposition of organic dye from wastewater. The green synthetized CuNPs were then characterized using UV–Visible Spectroscopy, Fourier Transform Infrared Spectroscopy (FTIR), Field Emission Scanning Electron Microscopy (FESEM), and X-ray Diffractometer (XRD). Further, tests were carried out in a batch system to evaluate the effect of initial dye concentration and contact time on methylene blue (MB) dye removal efficiency. The green synthesized CuNPs have effective photocatalytic activity against methylene blue in sunlight. The present research work of green synthesis of CuNPs using aqueous leaf extract and its further application for removal of hazardous dye from industrial wastewater provides promising approach because of its low-costs, low energy requirements and thus contributes to protect the environment as well as health by avoiding the use of toxic chemicals.

2 Materials and Methods

2.1 Chemicals and Materials

Copper Sulfate ($CuSO_4 \cdot 5H_2O$) and Methylene Blue dye (MB chemical formula $C_{16}H_{18}ClN_3S$, molecular weight = 319.85 g/mol, λ_{max} = 650 nm) of AR Grade was procured from SRL Chemicals and used without any further purification. Distilled water was employed for preparation of leaf extract, solutions of metal salt, and MB solutions.

Fresh *Azadirachta indica* leaves were collected from the campus of MIT-WPU, Pune, India in the month of September. In order to remove the dust and other debris, the leaves of *Azadirachta indica* were washed thoroughly using distilled water. Washed leaves were then dried in a hot air oven at 60 °C temperature for 120–180 min. The dried *Azadirachta indica* leaves were then grounded into fine powder for utilization in further experiments.

2.2 Preparation of Aqueous Extract of Azadirachta Indica Leaf

An aqueous extract of *Azadirachta indica* leaf was used for green synthesis of Cu-NPs. To prepare the aqueous extract, 20 g of dried *Azadirachta indica* leaves powder were added to 200 ml of distilled water in 250 ml 3 neck round bottom flask. The mixture was continuously stirred using a magnetic stirrer and heated at 60 °C for 60–80 min in oil bath.

The prepared extract of *Azadirachta indica* leaves was cooled down to room temperature and was filtered using Whatman filter Paper No. 42 to separate the leaf debris. The filtrate called dark pale-yellow color aqueous extract was used immediately or stored in refrigerator at 4 °C for further use. The prepared extract was used within one week of preparation for nanoparticles synthesis.

2.3 Synthesis of Copper Nanoparticles

In this study, copper nanoparticles were synthesized through the green reduction approach utilizing *Azadirachta indica* leaf extract which acts as primary reducing and capping agent. For the synthesis of copper nanoparticles, 200 ml of 0.05 M aqueous solution of $CuSO_4 \cdot 5H_2O$ was prepared which was then heated at 80 °C with magnetic stirring at 500 rpm. Afterwards, 50 ml of aqueous *Azadirachta indica* leaf extract was added to the 200 ml of copper sulfate solution in a dropwise manner with constant stirring and heating for one hour.

Fig. 2 Schematic representation of green synthesis of CuNPs

After complete addition of leaf extract, the resulting mixture was cooled down from 80 °C to room temperature. During the reaction, the color of the copper salt solution changed from bluish to yellowish to dark green, indicated the formation of CuNPs. The resulting dark color solution was centrifuged for 20 min at 7,000 rpm. Finally, pellets (dark brown color powder) were collected and was calcined in muffle furnace overnight at 40 °C. It was then stored at room temperature for further use. The schematic representation of green synthesis of CuNPs using neem leaf extract is as shown in Fig. 2.

2.4 *Characterization of Synthesized Nanoparticles*

The bioreduction of copper ions to CuNPs was initially confirmed by visual color change along with UV–Visible analysis (using UV–Visible Spectrophotometer Model UV-1800 by Shimadzu). The UV–Visible spectra for CuNPs confirmation and photocatalytic reduction of MB dye were recorded at a range of 200–800 nm. For FTIR analysis (Bruker, Model ALPHA II available at Central Instrumentation Facility, Savitribai Phule Pune University, Pune), CuNPs suspension was examined in a range of 500–4000 cm^{-1}. The functional groups of active components of neem leaves extract responsible for reduction and capping of synthesized copper nanoparticles were determined from this FTIR spectrum. For determining the structure and size of synthesized CuNPs were examined under Field Emission Scanning Electron Microscopy (FESEM, Model FEI Nova NanoSEM 450 available at Central Instrumentation Facility, Savitribai Phule Pune University, Pune). The crystal structure of synthesized CuNPs was analyzed using X-ray diffractometer (Bruker-D8 Venture).

2.5 Photocatalytic Degradation of Organic Dye

Batch experiments were carried out to evaluate the photocatalytic activity of CuNPs for degradation of Methylene blue dye in presence of sunlight. Initially, stock solution of 10 mg/L of MB was prepared. In the first experimental run, 10 mg of the synthesized CuNPs was dispersed in 100 ml of 10 mg/L solution of methylene blue dye solution at pH value of 8.0 at ambient temperature.

Afterwards, the mixture was exposed to sunlight. At regular intervals of 10–15 min, 3 ml of sample was taken from the mixture. The progress of dye degradation reaction was examined using UV–Vis spectrophotometer with maximum absorbance around 650–665 nm and contact time of 120 min. Furthermore, initial concentration of MB was varied to 5 mg/L and 2.5 mg/L with the same dosage of photocatalyst (10 mg) to evaluate the effect of initial concentration of MB on dye removal efficiency of CuNPs.

3 Results and Discussion

3.1 UV–Visible Analysis of CuNPs

The formation of CuNPs was initially confirmed by the change in color from bluish to yellow to green upon the addition of aqueous extract of *Azadirachta indica* into aqueous copper sulfate solution and is presented in Fig. 3.

Further, UV–visible spectroscopy analysis was used to confirm the formation of CuNPs. Figure 4 shows the UV–visible spectrum of synthesized Cu-NPs using *Azadirachta indica* leaf extract. Spectrum clearly showed the characteristics absorption peak at 800 nm confirming the formation of copper nanoparticles. This absorption peak is also called as surface plasmon resonance (SPR) of CuNPs which is due to electron excitation in the conduction band of NPs surface. In addition, the absorption peaks between 200 and 400 nm is related to double bonds present in phenolic compound in extract. As per the literature study, when the wavelength of incident light exceeds the particle size diameter causes the transition of electrons from valence band to the conduction band and gives the absorption maxima at ~500–600 nm indicating the formation of CuNPs. This SPR band varies based on the size and shape of the NPs. Similar results of previous studies related to size variations of

Fig. 3 Confirmation of formation of CuNPs by addition of aqueous neem extract to $CuSO_4$ solution

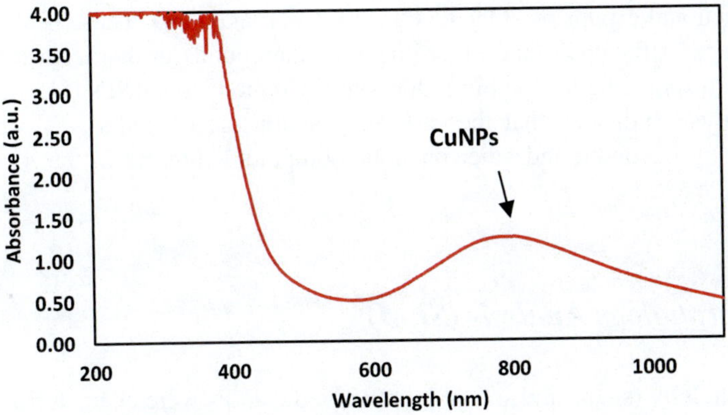

Fig. 4 UV–visible spectrum of synthesized Cu-NPs using neem leaf extract

green synthesized CuNPs and SPR band support the findings of the present study [3, 11, 14, 15].

3.2 *FTIR Analysis*

FTIR analysis was carried out to identify the biomolecules present in neem leaf extract which are responsible for reduction and capping of CuNPs. The FTIR spectrum of green synthesized CuNPs suspension exhibited in Fig. 5 with absorptions peaks at 3263.51, 1635.77, 3000, 1098.97, 1260.84 cm^{-1}, 941.91–506.33 cm^{-1} correspond

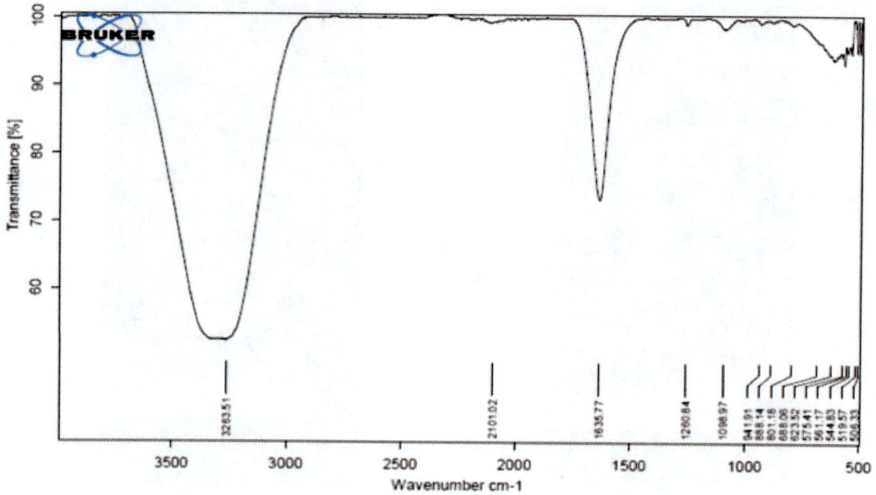

Fig. 5 FTIR Spectra of Copper Nanoparticles solution

to the functional groups of –OH, C=C, C–H, C–N, and C–O. These peaks indicate the presence of flavonoid and other phenolic compounds in the neem leaf extract which are responsible for the bio-reduction of Cu^+ ions to CuNPs [3]. Thus, FTIR result analysis indicated that the surface of synthesized CuNPs was capped and stabilized by flavonoid and other phenolic compounds present in the *Azadirachta indica* leaf extract.

3.3 Morphology Analysis (SEM)

The morphology (shape and size) of synthesized CuNPs were characterized by FE-SEM. As shown SEM image in Fig. 6, synthesized CuNPs from leaf extract have spherical-shaped morphology which was consistent with previous findings [13]. Slight agglomeration was observed in the synthesized copper nanoparticles. The size of the particles was found to be in the range of 50–70 nm.

3.4 X-ray Diffraction Analysis

To analyze the crystal structure and phase identification of synthesized copper nanoparticles sample, XRD analysis was performed and displayed in Fig. 7. Three major peaks appeared at 2 θ values of 43.48°, 50.40°, and 74.54° which belongs to (111), (200), and (220) planes of face centered cubic crystal structure of copper. These XRD results are closely matched with the literature values of metallic copper

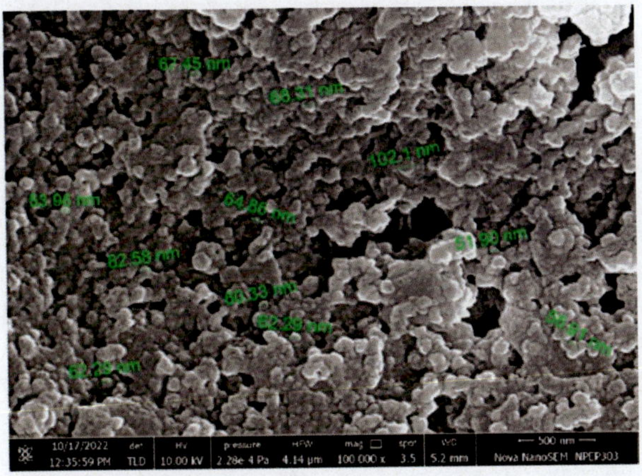

Fig. 6 SEM image of green synthesized CuNPs

[5]. In addition, the XRD pattern consists of small peaks (e.g. 36.50°, 47.5°) which indicates presence of small impurities.

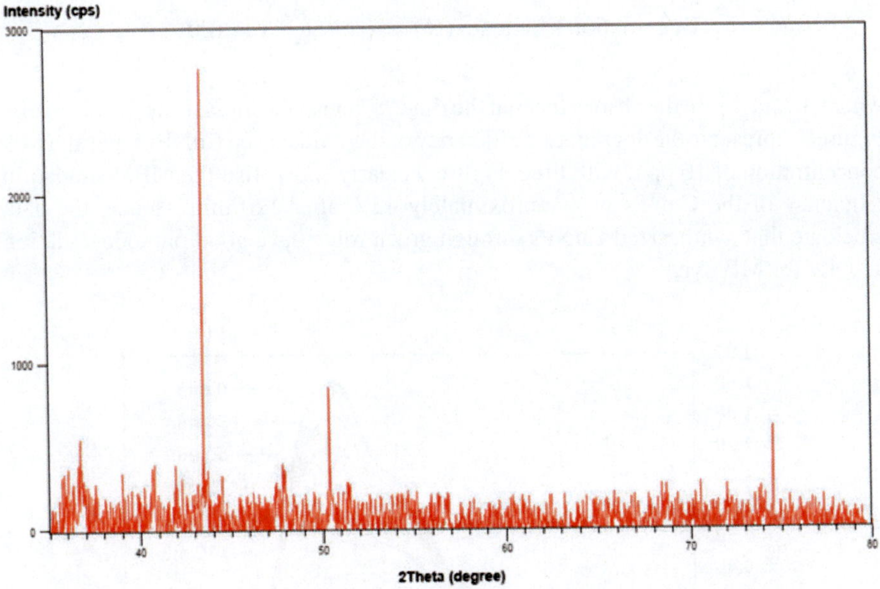

Fig. 7 XRD pattern of synthesized CuNPs

3.5 Photocatalytic Degradation of Methylene Blue Dye

The photocatalytic activity of synthesized CuNPs for degradation of MB was evaluated under direct sunlight at different experimental conditions as initial concentration of MB (10, 5, and 2.5 ppm) and contact time (10–120 min). The photocatalytic performance of CuNPs was analyzed using the UV–Visible spectra of MB samples within the wavelength range of 400–800 nm with maximum absorption around 665 nm.

Figure 8 shows the UV–Visible spectra for the first case of 10 ppm initial concentration of MB at different irradiation time. The recorded spectrum exhibits the characteristic band of MB at λ_{max} of ~650–664 nm (Fig. 8). It was observed that as the exposure time increased, the absorption peak corresponding to MB decreased gradually. Results clearly indicating the presence of CuNPs reduces the intensity of the MB and thus, its photocatalytic activity. As a result, color of MB gradually lightened (from dark blue to bright blue). In the absence of CuNPs, MB aqueous solution was kept in sunlight for more than two hours, resulted in negligible change in color and dye degradation. These experimental findings exhibited that dye degradation was driven by a photocatalytic activity of CuNPs. The efficiency of the CuNPs for the degradation of MB can be calculated using the following equation [8].

$$\text{Degradation Efficiency}(\%) = \left(1 - \frac{A_t}{A_0}\right) \times 100$$

where A_t and A_0 are the absorbances at the time "t" and at the initial time, respectively. Figure 9 represents the degradation efficiency of the synthesized CuNPs for initial MB concentration of 10 ppm with time. Figure 9 clearly shows that the MB degradation efficiency of the CuNPs was approximately 42% after 120 min. Hence, the data elucidate that synthesized CuNPs through green route have good photodegradation activity for MB dye.

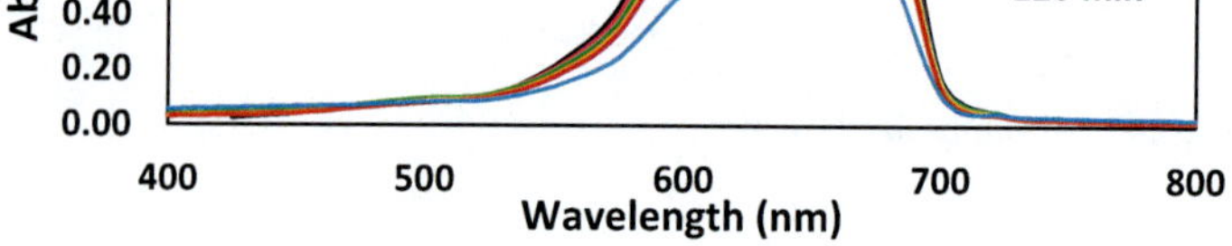

Fig. 8 UV–visible spectra for the reduction of MB (initial conc. of 10 ppm) using CuNPs at different time intervals

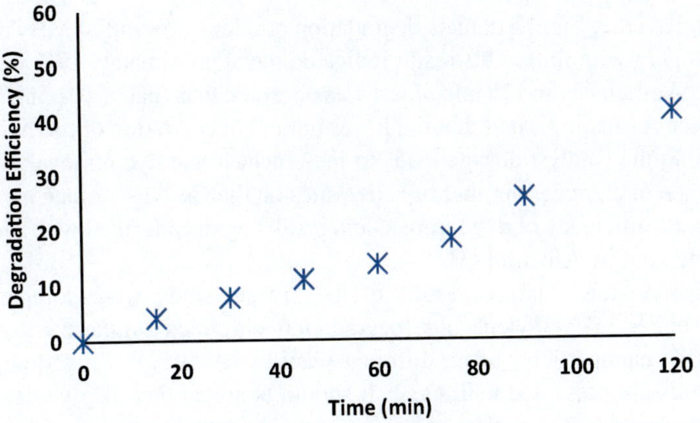

Fig. 9 % degradation efficiency of MB dye (10 ppm) with time

In general, steps in the degradation of methylene blue in presence of CuNPs and sunlight are as follows [3, 5, 8]. In first step, when the solar irradiation is absorbed by CuNPs, it gets photo excited due to SPR effect (Eq. 1). Secondly, the electron and holes generated in Eq. 1 can react with O_2 (Eq. 2) and H_2O (Eq. 3) molecules to provide active species, namely, hydroxyl radical (OH^-), and anionic superoxide radical (O_2^-), respectively (Eq. 4). Finally, both oxidation and reduction proceed on the surface of the photocatalyst Eqs. 5 and 6.

$$Cu + h\nu \rightarrow e^-(Cu) + h^+(Cu) \tag{1}$$

$$Cu\left(e^-\right) + O_2 \rightarrow O_2^- \tag{2}$$

$$Cu\left(h^+\right) + H_2O \rightarrow OH^- \tag{3}$$

$$O_2^- \text{ or } OH^- + \text{MB Dye} \rightarrow \text{degraded product} \tag{4}$$

$$MB + e^- \rightarrow \text{Reduction products} \tag{5}$$

$$MB + h^+ \rightarrow \text{oxidation products} \tag{6}$$

Later, similar experimental runs were conducted with initial concentration of MB of 5 and 2.5 ppm with same dose of CuNPs and results of degradation process are exhibited in Figs. 10 and 11. It is clearly observed in Figs. 10 and 11 that with decrease in initial MB concentration leads to an increase in the rate of dye degradation when concentration of nanoparticles remained the same. Obtained percentage degradation efficiency of MB at 5 and 2.5 ppm is approximately 50% and 60% respectively. As a representative case, Fig. 12 depicts degradation efficiency for initial MB concentration of 2.5 ppm with time. This result indicated the approximately 19% increase in MB removal efficiency in 120 min of test was observed than that of 10 ppm. Comparison of these results indicated that the lower initial concentration of methylene blue or increasing the catalyst dosage leads to increment in the dye removal efficiency. The presence of a greater number of active sites on the Cu-NPs surface is related to this increased efficiency of dye removal and results were confirmed with the similar study carried out by Alhalili [13].

Furthermore, the obtained results of the present study were compared with previous studies of methylene blue degradation via green synthesized copper or copper oxide nanoparticles using different plant extracts [5, 9, 11, 15] and degradation results are presented in Table 1. It should be noted that the dye degradation efficiency is dependent on various process parameters such as active elements of plant

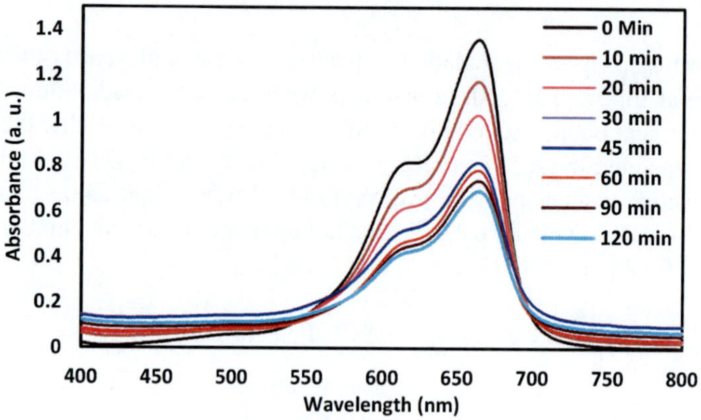

Fig. 10 UV–visible spectrum of MB degradation with irradiation time (initial MB Conc. 5 ppm)

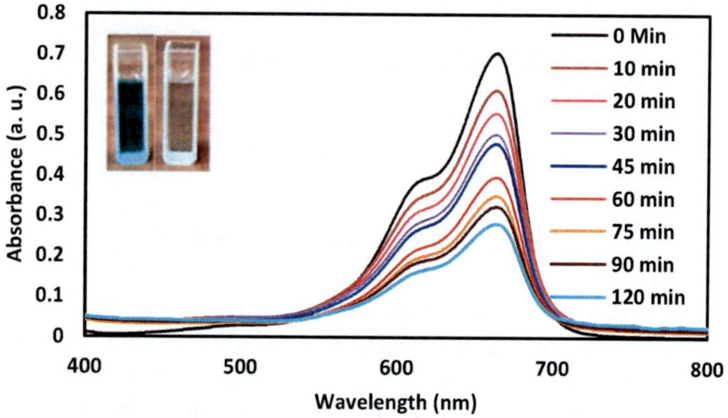

Fig. 11 UV–visible spectrum of MB degradation with irradiation time (initial MB Conc. 2.5 ppm)

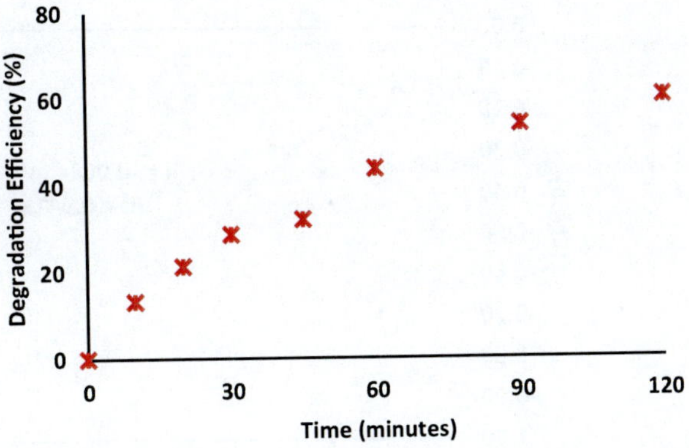

Fig. 12 Degradation efficiency of MB dye (2.5 ppm) with time

extract, concentration of dye, dosage of nanoparticles, sunlight or UV light exposure and irradiation time. Although there are different possible factors that affects the photodegradation process, a comparative study clearly illustrates that our results exhibited the similar trend of dye degradation (refer Figs. 8, 9, 10, 11 and 12]. However, similar type of study using aqueous extract of *Azadirachta indica leaves* has been carried out by Ref. [11] and our findings are in close agreement with their results.

The kinetics of the photocatalyzed degradation of dye was described by a pseudo-first-order reaction for the concentration of MB [3].

$$\ln\frac{C_t}{C_o} = -kt$$

where C_o is the initial MB concentration and C_t is the MB concentration at the irradiation time (t) and k is the rate constant (min^{-1}). The relationship between $\ln C_t/C_o$ versus irradiation time (t) presented in Fig. 13 showed good linear correlation with the

Table 1 Comparison of MB photodegradation via CuNPs synthesized using different plant extracts

Plant extract	Process conditions for dye degradation	Degradation efficiency (%)	References
Aqueous extract of *Azadirachta indica leaves*	20 ml of 10 ppm MB dye + 480 and 320 μl CuO NP solution + UV light + 25 min	45–58	[11]
Aqueous extract of fish scales of *Labeo rohita*	200 ml of 10^{-4} M solution + 10 mg Cu NPs + sunlight + 135 min	96	[5]
Aqueous extract of *Jatropha curcas* leaves	20 mL MB + 10 mg of the CuNPs + sunlight + 85 min	60–70	[9]
Aqueous extracts of leaf, stem and callus of *M. frondosa*	100 ml of 5 ppm dye + 20 mg of CuO-NPs + UV light + 100, 120 and 140 min	88–91	[15]

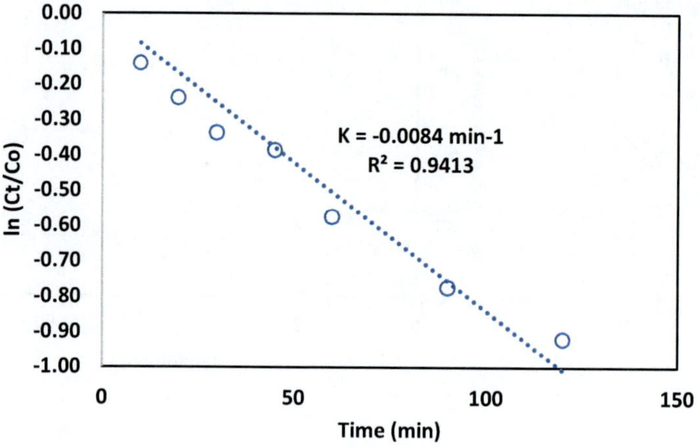

Fig. 13 Kinetic data for the MB degradation under sunlight irradiation

correlation coefficient of 0.94 ($R^2 > 0.94$). The slope of the linear fitting line as shown in Fig. 13 depicts the rate constant (k) of the reaction was found 0.0084 min^{-1}. From this kinetic study, it is concluded that the degradation of MB dye process resulted in pseudo-first-order kinetics with obtained value of rate constant 0.0084 min^{-1}.

4 Conclusions

The copper nanoparticles (CuNPs) was successfully synthesized using leaf extract of *Azadirachta indica* (neem) which is simpler, green, environment-friendly, economical method and free of the use of any organic/toxic solvents, and specialized instruments. The UV–Visible Spectroscopy and FESEM analysis results confirmed the formation of spherical CuNPs with average size in the range of (50–70 nm). FTIR analysis confirmed the presence of various functional groups responsible for the reduction and stabilization of CuNPs. The results revealed that 60% of hazardous organic methylene blue dye was completely degraded using green synthesized CuNPs under sunlight and thus, showed good photocatalytic activity. The developed method provides an environmental and sustainable process for synthesis of CuNPs and its utilization for removal of industrial hazardous dye. This experimental study can be extended for green synthesis of other metal nanoparticles using suitable leaf extracts and evaluation of their applicability in dye removal.

References

1. Lin, J., Luo, Z., Liu, J., Li, P.: Photocatalytic degradation of methylene blue in aqueous solution by using ZnO–SnO_2 nanocomposites. Mater. Sci. Semicond. Process. **87**, 24–31 (2018)
2. Hou, Y., Yan, S., Huang, G., Yang, Q., Huang, S., Cai, J.: Fabrication of N-doped carbons from waste bamboo shoot shell with high removal efficiency of organic dyes from water. Biores. Technol. **303**, 22939 (2020)
3. Mali, S.C., Dhaka, A., Githala, C.K., Trivedi, R.: Green synthesis of copper nanoparticles using celastrus paniculatus willd leaf extract and their photocatalytic and antifungal properties. Biotechnol. Rep. **27**, 1–9 (2020)
4. UN WWDR: WWAP (United Nations World Water Assessment Programme). The United Nations World Water Development Report 2017: Wastewater, The Untapped Resource. Paris, UNESCO (2017)
5. Sinha, T., Ahmaruzzaman, M.: Green synthesis of copper nanoparticles for the efficient removal (degradation) of dye from aqueous phase. Environ. Sci. Pollut. Res. **22**, 20092–20100 (2015)
6. Usman, M., Ahmed, A., Yu, B., Peng, Q., Shen, Y., Cong, H.: Photocatalytic pote ntial of bio-engineered copper nanoparticles synthesized from Ficus carica extract for the degradation of toxic organic dye from waste water: growth mechanism and study of parameter affecting the degradation performance. Mater. Res. Bull. **120**, 110583 (2019)
7. Khalafi, T., Buazar, F., Ghanemi, K.: Phycosynthesis and enhanced photocatalytic activity of zinc oxide nanoparticles toward organosulfur pollutants. Sci. Rep. **9**, 6866 (2019)
8. Mehrabian, M., Esteki, Z.: Degradation of methylene blue by photocatalysis of copper assisted ZnS nanoparticle thin films. Optik **130**, 1168–1172 (2017)
9. Ghosh, M.K., Sahu, S., Gupta, I., Ghorai, T.K.: Green synthesis of copper nanoparticles from an extract of jatropha curcas leaves: characterization, optical properties, CT-DNA binding and photocatalytic activity. RSC Adv. **10**, 22027–22035 (2020)
10. Nagar, N., Devra, V.: Green synthesis and characterization of copper nanoparticles using *Azadirachta indica* leaves. Mater. Chem. Phys. **213**, 44–51 (2018)
11. Rafique, M., Shaikh, A.J., Rasheed, R., Tahir, M.B., Gillani, S.S., Usman, A., Imran, M., Zakir, A., Khan, Z.U., Rabbani, F.: Aquatic biodegradation of methylene blue by copper oxide nanoparticles synthesized from *azadirachta indica* leaves extract. J. Inorg. Organomet. Polym Mater. **28**, 2455–2462 (2018)
12. Ansilin, S., Nair, J.K., Aswathy, C., Rama, V., Peter, J., Persis, J.J.: Green Synthesis and characterisation of copper oxide nanoparticles using *azadirachta indica* (Neem) leaf aqueous extract. J. Nanosci. Technol. **2**(5), 221–223 (2016)
13. Alhalili, Z.: Green synthesis of copper oxide nanoparticles CuO NPs from Eucalyptus Globoulus leaf extract: adsorption and design of experiments. Arab. J. Chem. **15**, 103739 (2022)
14. Amaliyah, S., Pangesti, D.P., Masruri, M., Sabarudin, A., Sumitro, S.B.: Green synthesis and characterization of copper nanoparticles using Piper retrofractum Vahl extract as bioredutor and capping agent. Heliyon **6**, e04636 (2020)
15. Manasa, D.J., Chandrashekar, K.R., Madhu Kumar, D.J., Niranjana, M., Navada, K.M.: Mussaenda frondosa L. mediated facile green synthesis of copper oxide nanoparticles–characterization, photocatalytic and their biological investigations. Arab. J. Chem. **14**, 103184 (2021)